Genetics In Practice

by F. B. Hutt

Author of Genetics of the Fowl

Norton Creek Press

http://www.nortoncreekpress.com

Genetics in Practice
Practical Chicken Breeding Methods

by F. B. Hutt
Norton Creek Press
36475 Norton Creek Rd.
Blodgett OR 97326
http://www.nortoncreekpress.com

Originally published in 1949 as Chapter 15 of *Genetics of the Fowl*.

Other Books From Norton Creek Press

See our Web site at https://www.nortoncreekpress.com.

- Success With Baby Chicks by Robert Plamondon
- Poultry Breeding and Management by James Dryden
- Feeding Poultry by G. F. Heuser
- Fresh-Air Poultry Houses by Prince T. Woods, M.D.
- Genetics of the Fowl by F. B. Hutt
- Turkey Management by Marsden and Martin
- Poultry Production by Leslie E. Card
- The Dollar Hen by Milo M. Hastings

Contents

Genetics in Practice

One of the author's former students confided (after he had safely graduated from the university) that, while genetics might have its good points, in his opinion it could be of practical value to the poultry breeder only if someone would separate the wheat from the chaff. Since some readers of this book may classify under that latter term most of the facts surveyed in the foregoing 14 chapters, it is perhaps desirable that a little wheat be put in this last one.

However, it should first be pointed out that in the previous pages it has been impossible to divorce facts from their applications. Consequently, any reader interested only in how to breed bigger and better chickens should at least glance through some of the other chapters for any kernels that may have been scattered, perhaps inadvertently, through them. This applies particularly to the discussion of egg production, in which the merits of various measures of productivity and periods for testing are reviewed, together with environmental influences that the breeder should consider and the relative importance of contributory variables that may influence the egg record for any hen and the flock to which she belongs. Similar reviews of many points that must be considered in breeding to increase body weight or size of egg are given in Chaps. 9 and 11. Procedures for eliminating lethal genes are outlined in *Genetics of the Fowl*, Chapter 8, and some of the problems involved in breeding birds resistant to disease are reviewed in *Genetics of the Fowl*, Chapter 12.

It is only with considerable misgiving that the author ventures to suggest how to improve our domestic fowls by breeding. One reason for his reluctance is that he knows from his own experience the inevitable disappointments, difficulties, and delays that will be encountered by the novice. Another is that many experienced poultry breeders have such firm convictions about methods to be followed that it is usually safer to listen to their expositions than to expound one's own ideas. Accordingly, the principles and practices given in that part of the ensuing discussion concerned with progeny testing are presented merely as a record of methods that the author has found to be successful. Others might be better.

For convenience, practical applications of genetics in poultry breeding will here be considered under these headings:

- Objectives and Methods
- Mass Selection
- Progeny Testing
- Inbreeding
- Utilization of Hybrid Vigor

Objectives and Methods

From the viewpoint of the geneticist trying to simplify matters, there are two general classes of objectives and two general methods of seeking to attain them. These are outlined in Table 1. They do not consider inbreeding and hybrid vigor, which are treated as special cases later on.

Table 1. Objectives of Poultry Breeders and Methods by Which Their Attainment May Be Attempted

Kind of objective	Examples	Method of breeding	Probable degree of success
Unifactorial characters	Rose comb. Recessive white. Any dominant allele of recessive lethal, or of other unifactorial defect. Blue eggshell.	Mass selection	Complete for recessive characters; incomplete for dominant ones (see Fig. 1)
		Progeny test	Complete
Multifactorial characters	High egg production. Size of egg, or of body. Resistance to disease. Conformation. Non-broodiness.	Mass selection	Effective up to a certain point, particularly in unimproved stock
		Progeny test	Permits progress beyond limits attained by mass selection

Unifactorial characters provide some of the distinguishing characteristics of breeds. Most of them show some variation because of multiple modifying genes such as those that cause the crest to range from the beautiful veil of the prize-winning Houdan down to the few stubby feathers of the geneticist's heterozygote. Similarly, barring is caused by a single gene in females, but variations in the width and straightness of the bars result from the action of an unknown number of modifiers. Other unifactorial characters are either present or not, with little evidence of modification, as, for example, recessive-white plumage or yellow skin.

Multifactorial characters generally show more variability, but their very elasticity is one of the things that intrigue the breeder and inspire him to see how far the species can be stretched. When, in 1913, Lady MacDuff was elevated to the peerage because she laid 303 eggs for James Dryden (1921), that first trap-nest record showing that a hen could lay over 300 eggs in a year provided for poultry breeders the world over an incentive by which they were literally egged on and on, until single records of that order became even too commonplace to advertise.

Hens not neglected in that respect include:
- The Barred Rock, Lady Victorine, that laid 358 eggs at Saskatoon in 1929.
- The well-named "No Drone No. 5 H," a White Leghorn with 357 eggs at the Agassiz laying contest in 1930.
- The 13 White Leghorns, 12 of them full sisters, that turned in an average production of 312.1 eggs per bird in 51 weeks at the western New York egg-laying test in 1945.

If not inspired to manipulate these multifactorial characters by the urge to produce superior stock, the breeder is driven to do so by economic necessity, for most of the things to which he looks for income depend upon an undetermined number of multiple genes. The raising of the standard weight for Leghorns in the United States was forced on the breeders by their recognition of the fact that bigger Leghorns lay more and bigger eggs.

Mass selection consists in mating together individuals of the type desired or which the breeder hopes will produce the kind of stock that he wants. It is based on the dictum, more familiar than accurate, that "like begets like." So far as simple recessive characters are concerned, mass selection is completely satisfactory. To illustrate, matings of recessive whites *inter se* yield only recessive whites; the same applies to single combs. Mass selection is less effective when unifactorial dominant characters are concerned, as will be made clear in the next section. Finally, it has serious limitations as a method for improving the multifactorial characters that are the chief concern of the modern poultry industry. Good layers do not always beget good layers, and hens hatched from big eggs may lay small ones.

The effectiveness of mass selection is enhanced somewhat if the breeding birds be evaluated, not only by their own appearance or performance, but also by the performance of their ancestors, *i.e.* by their pedigree. Especially to be desired in such a pedigree are progeny tests of the ancestors. At best, there is no assurance that an unproven cockerel will beget offspring to equal his parents. Full brothers with identical pedigrees frequently yield disappointingly different results.

Progeny testing means merely the evaluation of breeders according to the performance of their offspring. It rejects the notion that fine feathers make fine birds, and adopts instead the proverb "Handsome is as handsome does." In practice, it incorporates mass selection but goes one step farther. The good breeder selects his cockerels for testing only after careful consideration of their appearance, weights, and pedigrees (if available), but that is only the preliminary stage. The final evaluation and the selection of males for re-use in subsequent seasons depend upon careful evaluation of what their daughters show in egg production, viability, egg size, or whatever the objectives may be.

Included in a progeny-testing program is sib-testing. A sib-tested cockerel is one that is evaluated according to the performance of his brothers or sisters, or both. In that process one should consider the performance not only of his full sisters but also of his half sisters. The former provide a progeny test for the cockerel's dam, the latter for his sire. This does not mean that the sib-tested cockerel is progeny tested; he does not attain that rank until his own daughters have turned in their records. The important thing is that, both in sib tests and in progeny tests, the evaluation of the breeding stock is determined first by the record of the families to which they belong and finally by record of the families which they produce.

In genetic parlance, mass selection is selection according to the phenotype. By contrast, progeny testing is selection based on the genotype, which, in turn, is estimated from the kinds of offspring produced.

Mass Selection

For Unifactorial Characters

It is axiomatic that simple recessive characters must "breed true," except for accidental matings. They are maintained merely by excluding from the breeding pen those which do not show the desired type. The ease with which some simple recessive type may be extracted from a heterogeneous stock and developed into a uniform flock was nicely demonstrated by Lippincott (1920). He mated White Orpington males to as fine a collection of mongrel hens as one could hope to see and apparently found in the F1 generation no birds showing the recessive white of the Orpington. However, when some of the F1 females were backcrossed to another White Orpington, a good proportion of the resultant offspring showed, not only recessive white, but also fair Orpington type. Some of these birds which most resembled their sire were again backcrossed to a White Orpington male, with the result that their offspring, though only the third selected population, were uniformly recessive white.

It is more difficult to "fix" a desired dominant character. It is probable that Lippincott's recessive whites would have bred true for that character but would have continued for some years to throw offspring with yellow shanks merely because Ww heterozygotes cannot be distinguished from the true-breeding homozygotes of the genotype WW.

This point is illustrated by a problem that plagued the Wyandotte breeders for years, *i.e.,* the continuous outcropping of the despised single comb. In spite of its being systematically excluded from the breeding pens, good respectable rose-combed Wyandottes with no bar sinister in their pedigrees continued to produce single-combed chicks. This happens merely because the desired homozygote is indistinguishable from the heterozygote. When two of the latter, *Rr*, are mated together, the proportion of single combs in their offspring is about 25 per cent, but among the rose-combed chicks the proportion of heterozygotes is two-thirds. This corresponds to an F2 ratio. If one calculates the expectations in successive generations, assuming that all single combs are discarded but that all rose combs are mated at random and contribute equally to the next generation, it will be found that the proportions of the three genotypes in any ensuing generation, *n,* are given by the formula

$$F_n = (n-1)^2 RR + 2(n-1)Rr + 1rr$$

The rate at which the proportion of single combs decreases in successive generations, as calculated by this formula, is shown in Fig. 1. In the F_{10} generation, the proportions of the three genotypes are 81 *RR*:18 *Rr* 1 rr. Expressed otherwise, about one in every five or six rose-combed birds will still carry the gene for single comb even after 10 years of mass selection against it.

It is little wonder that mass selection is a slow method for eliminating undesirable recessives. The same persistence of the recessive allele would occur with any simple recessive character. With those which are lethal *(e.g.,* congenital loco, talpid, etc.) the recessive homozygote is self-eliminating, but the heterozygotes maintain the gene in the flock.

Multifactorial Characters

It is sometimes stated that the futility of mass selection was demonstrated in the Barred Rocks at the Maine Experiment

Station, when 9 years of that practice failed to raise their mean winter egg production. This failure was accentuated by the remarkable increase achieved by Pearl (1911) in his first year of progeny testing in the same flock, an increase which was maintained thereafter (Fig. 1). When these findings were coupled with the earlier demonstration by Pearl and Surface (1909) that daughters of 200-egg dams laid no more eggs than pullets from dams with records of 150 to 200 eggs, no justification for mass selection could be seen at Orono, Maine.

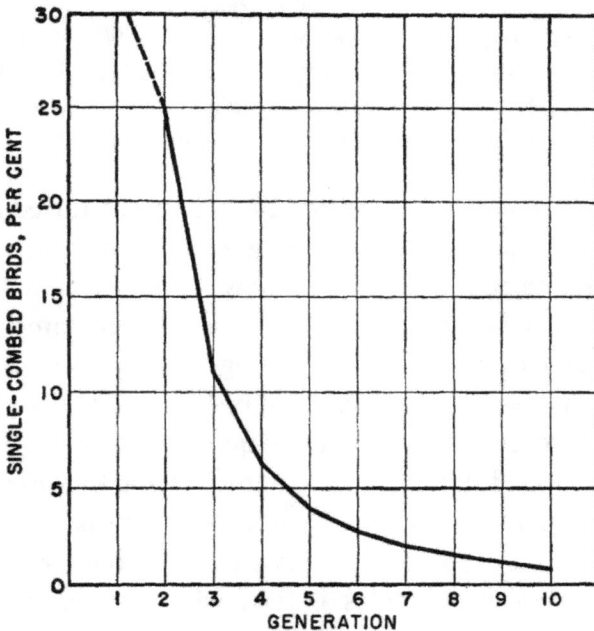

FIG. 1. Decrease of single-combed birds in successive generations from two heterozygous rose-combed parents, when only rose-combed birds are used for begetting the next generation. The rate of decrease is the same for any simple recessive, self-eliminating defect, such as a lethal gene.

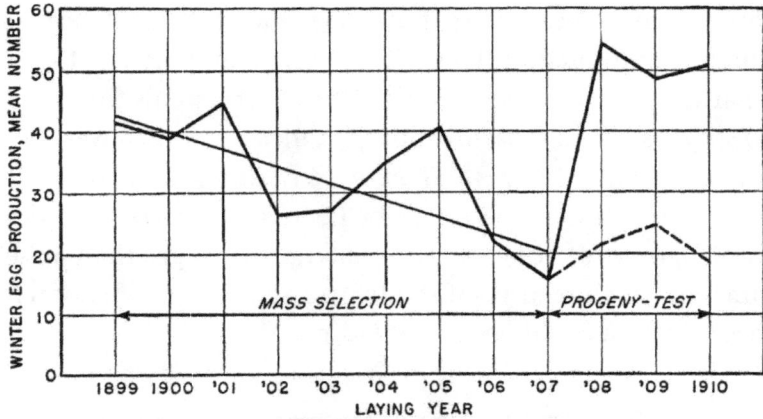

FIG. 2. *Effect of progeny testing on winter egg production. From 1899 to 1907, mass selection for high fecundity was apparently ineffective, but when progeny-tested birds were used, the winter production was immediately raised to a higher figure than in the previous 9 years. (From Pearl.)*

The fact remains that egg production can be raised by mass selection. At the very time when the spate of bulletins and reports deprecating that procedure was pouring from the Maine station, at the other side of the continent James Dryden (1921) was demonstrating at the Oregon Experiment Station, not only that egg production could be increased by mass selection, but also that it could be done quickly. This he did with no less than three different breeds, one of which, the Oregons, he himself originated. A summary of these results is shown in Fig. 4. Dryden selected breeding hens according to their individual trap-nest records and males on the basis of the records of their dams. In 10 years, the mean egg production was raised from 86 to 215 in Barred Rocks, from 135 to 231 in Oregons, and from 107 to 212 in White Leghorns.

FIG. 3. Raymond Pearl, who contributed much to knowledge of the physiology of reproduction in the fowl and early stressed the importance of the progeny test. (Courtesy of Mrs. Pearl.)

The differing results shown in Figs. 2 and 4 are perhaps not so contradictory as they seem. It is evident from the graphs in Fig. 4 that Dryden's greatest progress was made in the earlier years of selection, and that no improvement (in the flock averages) was made after 1914 in the Leghorns or after 1915 in the Oregons. With the Barred Rocks, progress was slower than in the other two breeds but more consistent. Obviously, mass selection is most effective in an unimproved stock but less so after a certain level has been attained. In fowls that level may be around 200 to 220 eggs. It is not surprising, therefore, that Pearl should have found little merit in mass selection. The fact that Pearl and Surface (1909) were able to house in the fall of 1907 no less than 250 pullets from 200-egg hens (and 600 more from dams with records of 150 to 200 eggs) suggests that by that time their predecessor, Professor Gowell, had already improved the stock to the point beyond which little progress was likely to be made by mass selection.

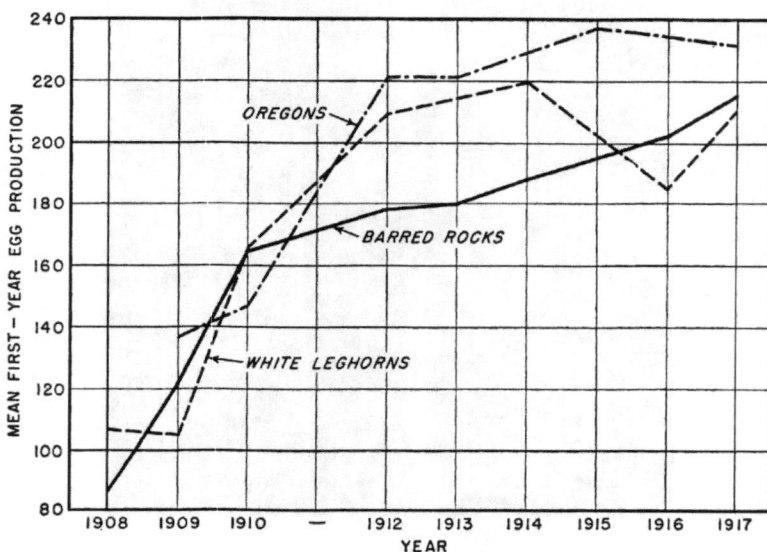

FIG. 4. Increase of egg production by mass selection at the Oregon Experiment Station. The report indicates that no birds were raised in 1911. (From data of Dryden in Oregon Agr. Expt. Sta. Bull. 180.)

This interpretation does not detract in any way from Pearl's valuable demonstration that, when mass selection has ceased to be effective, remarkable progress can still be made by progeny testing. No one recognized this fact better than Dryden, who, without discarding mass selection as a useful procedure, wrote that "More rapid progress will be made by the breeder if he can test the breeding quality of his stock, and use for breeding those hens and males whose progeny has shown high production."

The limitations of mass selection and the efficacy of progeny testing have also been demonstrated at the other end of the scale of egg production. During the 22 years from 1913 to 1935, mass selection was practiced at Cornell University to determine the feasibility of thus raising productivity in one strain of White Leghorns and of lowering it in another. Results in the low-fecundity line, as reviewed by Hall (1934) and by Lamoreux *et al.* (1943), showed no significant change in 22 years, at the end of which time mean egg production was approximately the same, 100 eggs per hen, that it had been in 1913. Beginning in 1935, mass selection was discarded, and by 5 years of progeny testing the egg production was lowered to 40 eggs per bird (Fig. 6).

FIG. 5. James Dryden, a leader in the demonstration of fundamental princi-ples applicable in poultry breeding. (Courtesy of Horace Dryden.)

It should be noted in Fig. 6 that in the high-fecundity strain, mean egg production was steadily increased to a little over 200 eggs in 1928, after which time it remained unchanged for the next 4 years. As Petrov (1935) pointed out, some of the increase in productivity during the long course of the mass selection must be attributed to improvements in management.

The effect of one of these—provision of greater feeding space at the mash hoppers—is very obvious in Fig. 6. Nevertheless, this experiment tends to substantiate evidence from the other experiments cited above that mass selection will raise productivity of unimproved stock but has little efficacy in flocks with average production around 200 eggs.

Although this exposition of the possibilities and limitations of mass selection deals only with egg production, the principles apply equally well to other multifactorial characters, not only in the fowl, but also in other species. Obviously, mass selection will be more effective with characters in which the phenotype gives a good clue to the genotype than when these two are less closely related. For that reason it should be more effective in selection for body size, or for conformation, than in selection for ability to lay eggs. Many questions remain unanswered. For example, if, after some years of using the best 30 per cent of the flock as breeders, one reaches a limit beyond which further progress seems unlikely, can that limit then be raised by using only the best 10 per cent? If so, by how much? For these questions, as for many others, the author would like to know the answers.

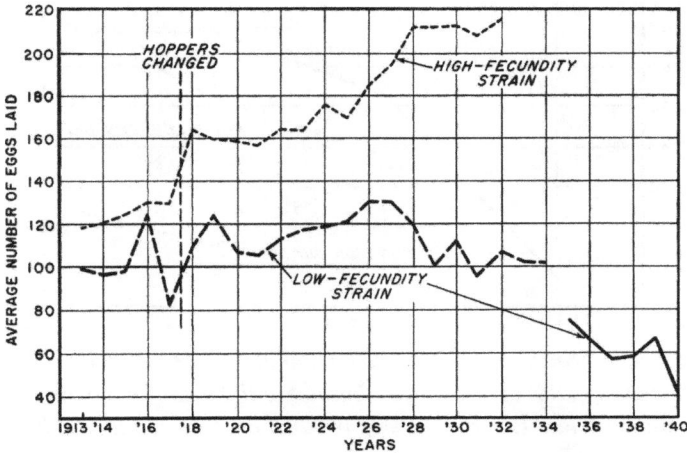

FIG. 6. Average egg production of two strains of White Leghorns selected for low or for high fecundity by mass selection at Cornell University from 1913 to 1934. In the high-fecundity strain, production was raised to a little over 200 eggs, after which point it remained fairly constant. In the low-fecundity line, production in 1934 remained about the same as in 1913, but it was significantly lowered thereafter by progeny testing, which was begun in 1935. (From Lamoreux et al. in Poultry Sci.)

Special Value of Mass Selection in Grading

The consistent use of superior sires, a process known as "grading," is generally considered by breeders of the larger animals as one of the most effective ways of improving unimproved stock. It is both rapid and comparatively inexpensive. Grading has not found much favor with poultrymen in the United States, perhaps merely because it is so easy and cheap to obtain purebred stock. Nevertheless, for the many great areas of the world in which the native fowls have not been improved, grading is the most practicable method of improving egg production, egg size, and general uniformity. It is particularly to be recommended because the supposedly unimproved native fowls have in reality their own desirable contribution to make, i.e., the resistance to disease bred in them by natural selection.

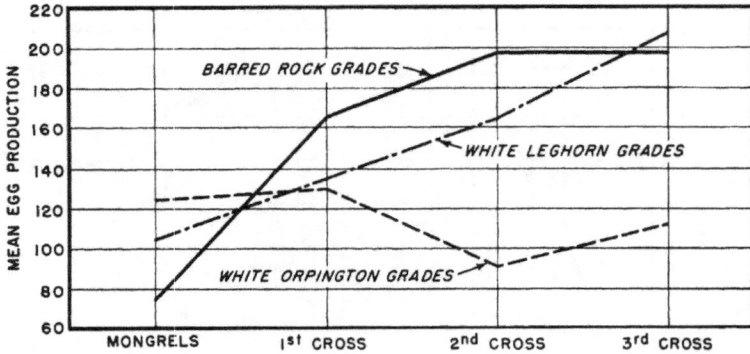

FIG. 7. *The improvement of egg production in 3 years by grading. This was effective with males of two breeds, but not with those of a third, in which there had been less selection for egg production. (From data of Lippincott.)*

The demonstration by Lippincott (1920) of the remarkable improvement in egg production that can be made by grading, while considered in his own country to be of interest more academic than practical, has a special significance for anyone seeking to increase the productivity of native stocks anywhere. In this case, mongrel fowls and their offspring were back-crossed for 3 years to purebred males of three different breeds and all from strains bred for egg production. None of these males was a proven sire.

The results (Fig. 7) are of special interest because during the 3 years the female breeders were selected merely to resemble the pure breed of their sires, and not with any consideration either of their own egg records or of the production of their dams. This was therefore mass selection with respect to the sires only. With the Barred Rock and Leghorn males, average production was raised in 3 years to 207 and 198 eggs, respectively. The failure of the Orpington males to effect any improvement was attributed to their having behind them less breeding for egg production than had the males of the other two breeds.

The effectiveness of grading in this way is well known to breeders who have used it in various parts of Africa, India, and the Malay Peninsula, but other records as complete as those of Lippincott are unknown to the author. Apart from egg production, it was clear from his illustration that continued back-crossing to one pure breed, coupled with some selection of the females for the desired type, will quickly convert any mongrel population into a flock with remarkable uniformity both in color and in conformation.

The Multiplying Million

In further justification of mass selection, it is well to remember that the progeny testers comprise only a small fraction of 1 per cent of the poultry breeders. In the United States over 90 per cent of the chickens raised each year come from the farm flocks supplying the commercial hatcheries. It is doubtful that one egg in a million passing through such hatcheries comes from proven dams or sires. Nevertheless, the quality of the chick has been steadily improving over the years, and it should continue to do so.

This is partly because progressive hatcherymen ensure that their supplying flocks are headed by the best cockerels that they can get from the breeders who do the progeny testing. This process provides a steady distribution to the farm flocks of the desirable genes accumulated by the good breeders. The numerically insignificant progeny testers thus bear the responsibility for the uplifting of the masses, but they share it with the hatchery-man and with the multipliers, the farmers who supply in millions of eggs for hatching the genes upon which a profitable poultry industry depends. These multipliers have their own responsibility. It is concerned almost entirely with selection of the hens that produce the eggs. These must be regularly and thoroughly culled to ensure that only the most desirable females are used for reproduction. The undersized, the late-maturing, the early-moulting, and the low-producing hens all must go. So must the small or otherwise undesirable eggs, even if the hen that lays them cannot be detected. All this is mass selection. It is important, and its contribution should not be overlooked. Neither should that of natural selection, which does the chick buyer a good service when it removes from the supplying flock the females most susceptible to lymphomatosis, to blackhead, or to any other ailment.

Progeny Testing

The Unreliability of Pedigrees

After upholding thus the merits of mass selection, it is appropriate to turn again to consideration of its limitations. While these are fairly evident in Figs. 2, 4, and 6, they are made still clearer if illustrated by specific cases in which the pedigree of the breeding bird has proved a false guide to the inheritance transmitted by it. Examples of this situation are all too familiar to the poultryman who pays for breeding cockerels according to the egg records of their dams. Table 2, which gives the records for 15 White Leghorn males used in 1 year at Cornell, shows what often happens in such cases.

Table 2. Progeny Tests of 15 White Leghorn Males, Showing Lack of Relationship between Egg Production of the Male's Dam and That of His Daughters

Egg production of male's dam in 365 days	Males, number	Mean egg production of daughters to 500 days of age
218-224	2	141
244-250	3	169
250-260	2	176
264-265	3	144
277	3	177
296-299	2	132

These records *do not* show that sons of 296- and 299-egg dams produce poorer layers than sons of hens that lay only 218 or 224 eggs. They do show merely that, in dealing with records above 200 eggs for first-year production, the performance of the sire's dam does not give much assurance of how many eggs his daughters will lay. One should not expect differences of 20 to 40 eggs in the records for dams to be reflected in the transmitting abilities of their sons; but when larger differences than these are involved, the egg production of the sire's dam does indicate—in general terms—what may be expected in his progeny. As Hall (1935) has shown, this is evident when large numbers of sires are grouped according to differences of 50 eggs in the records of their dams. Most progressive breeders are concerned nowadays only with testing cockerels from hens that have laid upward of 200, 250, or even 300 eggs. In that range of egg production, cases in which the sire does not transmit fecundity proportional to his dam's record or to records of several ancestors (Jull, 1934) are numerous enough to make mass selection unreliable and hence to necessitate progeny testing. Pride of pedigree is no guarantee that one may also take pride in the progeny.

Similarly, when only good dams of this type are concerned, any relationship between the record of dam and that of her daughters is not very conspicuous. This is too obvious in any breeder's records to need further amplification here, but it does not mean that selection is ineffective. It is fashionable, nowadays, to prove by abstruse statistical procedures that variations in egg production (and in other multifactorial characters of domestic animals) are influenced more by environment than by heredity. Breeders should not conclude therefrom that selection for the improvement of such characters is futile and that they should concentrate on the environment. While the environmental influences may completely outweigh genetic influences in any one year or in every year, the fact remains that the genetic variations can be accumulated by selection, whereas those caused by the environment cannot. Obviously, it is essential for the breeder to provide the environment most conducive to full manifestation of genetic differences in the stock.

The unreliability of pedigrees is perhaps best illustrated by the fact that full brothers (or full sisters) with identical pedigrees may produce families for which the records of performance are entirely different. Examples are given in the following section.

The Unreliability of Phenotypes

One desirable genetic character, for which the breeder is usually more willing to rely upon mass selection than upon progeny testing, is viability. It is true for hens, as for man, that the individuals which have demonstrated their fitness to survive are more likely to beget similarly disease-resistant offspring than are those which died early. The bearing of this fact on resistance to lymphomatosis was shown in *Genetics of the Fowl,* Table 57. It is also demonstrated by the comparatively high resistance to different diseases developed by natural selection in various parts of the world. Because of the conviction common to many poultrymen that old fowls must produce highly resistant progeny, it seems desirable to show that this does not always happen. Expressed otherwise, the phenotype with respect to viability may be entirely different from the genotype.

Figure 8 shows two White Leghorn males that lived to be used for three consecutive breeding seasons, one of them for four. Obviously, both demonstrated remarkable fitness to survive the rather severe exposure to lymphomatosis that is routine treatment for most Leghorns at Cornell. From their appearances one might prefer K 3450. However, as the data in Table 3 show, this male lost annually for 4 years from 60 to 76 per cent of his daughters. Of those alive at 6 weeks, from 33 to 47 per cent died of neoplasms (mostly lymphomatosis) before 500 days of age. By contrast, daughters of the other male, K 3459, experienced comparatively low mortality (for unculled birds deliberately exposed to disease) and almost negligible losses from neoplasms. Offspring of both sires were reared together.

FIG. 8. K 3450 (left) and K 3459 (right), two hardy males that were used for four and three breeding seasons, respectively. One of them produced offspring resistant to lymphomatosis and subject to comparatively low mortality. The other lost annually 60 to 76 per cent of his daughters. Table 3 identifies each.

Table 3. Comparison of the Two Males Shown in Figure 8 with Respect to Viability and Egg Production of Their Daughters.

Sire and year	Daughters alive at 42 days of age, number	Died 42-500 days, per cent		Eggs per bird,* number	Production index
		From all causes	From neoplasms		
K 3450 (susceptible):					
1943	55	60	32.7	163	102
1944	90	73	46.6	177	88
1945	216	64	41.2	151	93
1946	82	76	46.9	189	98
K 3459 (resistant):					
1943	56	27	3.6	210	191
1944	86	17	2.3	198	186
1945	69	25	11.6	213	186
* Average number laid to 500 days of age by those then alive.					

The remarkable difference in viability between the progenies of these two males resulted chiefly from the fact that one had been bred for susceptibility to lymphomatosis, the other for resistance to it. The genes contributed by susceptible dams in the one line and resistant dams in the other must also have influenced the viability of their daughters, but probably less than genes from the two sires. Both these were used for several consecutive years merely because they excelled contemporary males, the one in producing susceptible offspring, the other in begetting resistant stock. The important point is that, of two males surviving past 4 years of age, one may sire disease-resistant offspring and the other exactly the opposite. Only a progeny test can distinguish between the two.

The consistent performances of both these males with

respect to fecundity, viability, and production index of their daughters in successive years should be noted. This is only one of many records available to show that sires proven good, bad, or mediocre in one breeding season will transmit the same capacities in subsequent ones, except for genetic differences in the females with which they may be mated and except for environmental influences.

K 3459 was one of four full brothers mated in 1943 to four groups of comparable females. For three of these the mean egg production of their daughters was 183, 193, and 210, but for the fourth male the figure was only 166. All had the same pedigree; only the progeny test revealed the difference of 44 eggs per daughter between the worst and best of the brothers.

The impossibility of predicting the breeding performance of full brothers is also illustrated by the varying proportions of daughters with stubs (feathers on the feet) sired by 15 cockerels in one strain of White Leghorns at Cornell (Cole and Hutt, 1947). These comprised five trios of full brothers, and the members of each trio were used successively in the same breeding pen. None of these males had detectable stubs when used, nor had the 80 females in the five pens. All phenotypes were thus identical. So were the pedigrees of the brothers in any one pen, but remarkable differences in their genotypes were revealed by the progeny tests (Table 4).

Stubs are caused by an undetermined number of multiple genes. Only in one series, J 3, did all three brothers transmit similar inheritances. Since the maternal contribution was reasonably constant in each pen (being affected only in so far as some hens produced more chicks by one brother than by another), and since the incidence of stubs is presumably influenced little or not at all by the environment, the record in Table 4 is a useful indication of the extent to which full brothers may differ in the inheritance transmitted to their offspring.

Table 4. *Differences between Full Brothers Mated to the Same Hens in the Proportions, per Cent, of Their Daughters Showing Stubs*

Breeding pen	First brother	Second brother	Third brother
J 1	17	29	8
J 2	44	4	54
J 3	34	29	29
J 4	28	20	38
J 5	36	7	45

Records for Progeny Testing

In its simplest form, the progeny test requires few records, or even none. Any mating of two Wyandottes that yields a single-combed chick is a progeny test revealing both parents to be heterozygous for rose comb. Any two birds producing any simple recessive lethal character have had their genotypes with respect to that character revealed by the progeny test in its simplest form.

With multifactorial characters that are not measurable until the birds are 6, 12, or 18 months old, the problem is more complicated. For a simple one, such as egg production, the following records are desirable, in addition to those necessary for pedigreeing all offspring and for maintaining trap-nest records in the laying pens:

1. *Individual record.* A cumulative record of the eggs laid by each daughter.
2. *Dam's summary.* A grouping of daughters by dams, with an average for the group, thus revealing the good, bad, and mediocre members in each family of full sisters.
3. *Sire's summary.* A grouping of all dams mated to any one sire, with averages for each dam's daughters and for all the sire's daughters, thus revealing which dams have raised the

average for that sire and which have lowered it.

4. *Comparison of comparable sires.* A grouping of all sires, with averages for daughters of each, and with a grand average for all, so that sires which have produced superior daughters are easily recognized.

Such a system of records is fairly simple in so far as any one character is concerned. It becomes more complicated for the breeder who wants to consider, not merely egg production, but also hatchability, mortality during rearing, mortality after housing, age at first egg, weight of egg, weight of birds, and freedom from assorted real or imaginary defects. However, poultry breeders as a class take a special pride and delight in devising just such records. Their ingenuity in condensing indispensable data into the irreducible minimum of space is surpassed only by the zeal and enthusiasm with which the minutiae of the most complex system are described to anyone who will read or listen.

Since the time required for maintaining the necessary records is one of the chief deterrents that prevent some poultrymen from progeny testing, it is essential that the data recorded and the analyses required should be simplified as much as possible. The procedure of determining familial differences in size of egg by weighing for only one week in March, and perhaps for another in December (see *Genetics of the Fowl, Chapter* 11), illustrates one way in which record keeping is simplified. It is also facilitated by measuring egg production only to 500 days of age (Chap. 10) rather than for 365 days from first egg. Using the median to determine familial differences in age at first egg (page 309) is a short cut that reduces by about one-half the record keeping necessary when that same variable is measured by the mean.

One of the big tasks is that of transferring the records for individual birds to those of the dams' families. Some breeders have greatly simplified this process by keeping the individual records on printed, punched cards. These can be quickly sorted according to sire families either by card-sorting machines, if the number of birds be large enough to warrant that expense, or manually, with the use of a sorting needle. Another simplification is to reduce to a minimum the number of times a year when the family averages are computed. The author does this four times, as follows:

1. *On January 1,* in order to find which of the males and females used in the previous breeding season seem most worthy of being used in the next one and to find the best families from which to select cockerels for testing.
2. *On April 1,* a reappraisal of families to determine from which breeders it would best pay to save cockerel chicks for the next year.
3. *On July* 1, another examination, primarily to decide which families and individuals to discard as soon as their 500-day test is completed. Since the birds hatched first reach that age about July 20, it is necessary to have the list of unwanted birds ready by that date.
4. *About September 20,* a final appraisal to select the best families and (from them) the best individuals to keep for breeding. This is made as soon as the birds hatched latest have finished their 500-day test.

The second of these summaries might be eliminated, especially if enough proven sires and dams are used in the breeding pens to provide all the cockerels that are to be kept. Cumulative egg records of the pullets need not be added to the dam's summary from the individual records except when family averages are to be computed. Sometimes the quarterly summaries are begun on Dec. 1 instead of Jan. 1, but the longer test provided by the latter date permits a better evaluation of families.

The Use of Such Records

Poultry breeders like to set up standards for their breeding stock. The usual thing is to say that they breed only from birds of a certain body weight, with egg records not lower than 200,

230, or 250 eggs (according to available resources) and with egg size not smaller than 2 ounces. Standards are perhaps indispensable for national schemes for poultry improvement. A more flexible measuring stick sometimes has its advantages, as, for example, when an outbreak of coryza or of Newcastle disease lowers the average production of the flock some 20 eggs or so below that of previous years.

A flexible method of evaluation is provided simply by comparing the records of individuals and of families with the averages for the appropriate populations. A single daughter is compared with the average for her full sisters; a dam's record is compared with the average for all dams mated to the same sire; any one sire's record is compared with the average for all contemporary sires. This method has several advantages. It can be used before the progeny test has progressed further than 3 or 4 months or at any period that may be desired before the full test is completed. The success with which this method of ranking the families reveals differences between sires in the transmission of egg size, and does so long before definitive egg size is attained, is shown in *Genetics of the Fowl*, page 366. The illustration is important because any breeder using progeny tests wishes to evaluate his families, to decide which sires to use again, and to pick his breeding cockerels by the time when breeding pens should be made up. In northern latitudes that time is usually early in January. This he can do with most multifactorial characters, even though his progeny tests are less than half completed by Jan. 1, by ranking his families in relation to the averages.

When evaluating the families according to their rank above or below the mean, one actually uses the same measure for egg production, egg size, viability, or any other objective, and it is somewhat easier, therefore, to detect the family that is superior in several different items. In other words, it is easier to think of a sire as ranking first in egg production, fifth in egg size, and second in viability than to remember that the average records to Jan. 1 for his daughters were 69.7 eggs, of 53.4 gm., with mortality 5 per cent since 6 weeks of age. Another reason for evaluating families according to their rank rather than by standards is that, when two or three "shifts" of males are used in one breeding season (page 520), the differing environments for the chickens hatched early, in mid-season, or late necessitate separate standards for each group. Comparisons within each series with the average for that series will solve the problem.

In selecting breeding stock by the use of such records, and considering for illustrative purposes only one objective, the first procedure is to exclude from consideration the sire families that are below average. Such sires should not be reused, nor would their offspring be desirable, but one might well reuse any dams in such sire families that have clearly produced superior progeny in spite of the poor contribution of the sire. The next step, working only with the superior sire families, is to exclude from these the dam families showing less than average performance. One is thus left with the families that have both superior sires and superior dams. From these should come the cockerels and any pullets that are to be used for breeding. Obviously the sires and dams of such families should be reused—if they are still available.

This does not mean that every above-average family is worth breeding from. The intensity of selection must be varied according to the number of breeding birds needed, the number of objectives, and their relative importance. Thus, if egg size is already fully satisfactory (and especially since beyond a certain point bigger eggs are undesirable), one may select families that are merely average in that respect, or even below it, in order thus to have more families, dams, sires, cockerels, or pullets among which to select for egg production, for viability, or for both. Similarly, if mortality should be no special problem, the breeder may be more concerned about numbers of eggs than anything else. In that case, he might use any families that are average or better in viability, but only the top 20 per cent with respect to egg production.

Another way of stating the basis for selection outlined above is that any cockerel selected for testing is evaluated by the performance (1) of his full sisters and also (2) of his half sisters. These two measures provide estimates of the genes contributed to the cockerel by his dam and his sire, respectively. The superiority of such an appraisal over one based only on contributions from the dam's side has been emphasized by Godfrey (1946).

Complications and Limitations

Lest any novice should assume from the foregoing paragraphs that science has reduced the esoteric art of poultry breeding to anything so commonplace as the law of averages, it is desirable to enumerate some of the difficulties that make progress slow in spite of the most assiduous efforts of the progeny tester. They can be summed up merely by saying that it is difficult to find many proven sires good enough to use for several years and to find them in time to do so.

These difficulties result chiefly from the fact that nowadays no breeder is satisfied with a single objective. If only egg production be considered, about half the sires are above average; but with each additional independent objective the chance of getting a sire superior in all is again halved. The chance that a sire will be superior with respect to the hatchability, egg production, egg size, and viability of his offspring is reduced to about 1 in 16. Even if such a sire be found, his superiority may not be recognized in time to ensure that he be used again in the second breeding season. By the third, he may be dead or reduced in fertility. Of 24 Leghorn cockerels tested by the author in 1943, 2 died early, and 7 of the remainder eventually proved superior both in egg production and in resistance to disease. Three of these were recognized in time to permit their use in 1944, but only 2 were still on hand for the breeding season of 1945.

This dismal record shows why progeny testing on a small scale is not likely to be very successful. Some breeders who have tried it with only half a dozen pens for single-male matings have become discouraged at their lack of progress. During eight generations, Hays (1940) was able to make no improvement in egg production by 8 years of selection in small flocks of 20 to 50 hens. Some breeders with ample breeding pens make little progress because they divide these resources among several strains or breeds and are therefore unable to test enough males in any one of them to find the outstanding sires upon which progress depends.

Another limitation may be lack of adequate laying pens in which to maintain the pullets that provide the tests. This applies particularly in breeding for resistance to disease. As Mueller and Hutt (1946) have shown, one needs more than 30 daughters to differentiate resistant and susceptible families when mortality is high (37 to 50 per cent) and at least 50 daughters when mortality is comparatively low. In such work, to test even 24 cockerels a year requires space for about 1,200 pullets in addition to the others hatched from proven sires and dams. Requirements of space and the task of the record keeping can be reduced by discarding:

1. All daughters of any cockerels that die before housing time. There is no point in testing a cockerel that can never be used.
2. All daughters in sire families too small to provide an adequate test.
3. Surplus daughters above the 50 to 60 needed per sire. Since some families will have large surpluses and others small ones, or none, the pullets eliminated must be picked at random in order not to prejudice the sample that remains. It will not do to eliminate all late-hatched pullets in one family if such birds are retained in another. To reduce to families

of 60, one could eliminate every third daughter from a family of 90, every fifth one from a family of 75.

4. Daughters of dams that have only one or two pullets, as long as there are plenty of other daughters from at least six different dams. These small families tell nothing about the dam and add little to the test of the sire, but they do complicate the records.

Tests of dams for viability of their offspring are not so satisfactory as tests of sires. This is partly because most dam families are too small. Another reason may be that one dam's family by a single sire is a less reliable indicator of genotype than one sire's family from a dozen or more dams. This latter difficulty is removed by diallel or polyallel crosses in which one dam is mated at different times with two or more different sires.

Smaller families than these will suffice to determine the laying capacity transmitted by sires and dams. Mueller and Hutt obtained some evidence that as few as 6 daughters with complete egg records may be adequate to test a dam in this respect. However, to have 6 daughters survive to the end of the testing period, it may be necessary to start at least 10 or 12.

It has usually been considered that no culling should be practiced in families under test, but Bird and Sinclair (1938) found that it is feasible to cull 15 to 25 per cent in comparatively large populations without disturbing the relationships among them of their means with respect to egg production. They considered that any culling in families under test should be no more than enough to eliminate the lowest producers, those which cause an abnormal "low tail" in the distribution curve. Obviously, any culling in families to be compared would have to be done at an equal rate in all of them. The feasibility of doing so in 12 different sire families without obscuring differences between them was shown by Lerner and Taylor (1940). They even considered that families thus culled could still be ranked with respect to viability, but the feasibility of doing this in flocks from which the birds are actually removed has yet to be demonstrated.

In addition to the complications considered in the foregoing paragraphs, others will occur to any breeder. Many of these are concerned with the environment, which is considered later. The limitations imposed by it are suggested by the fact that at least one breeder, after exhaustive analysis of his records and selection there from of the outstanding family of the year, found upon going to the rearing range to inspect the cockerels of that family that their superior merits had already been recognized by others, even without the records, and that the foxes had been there first.

Double Shifts, Triple Shifts, and Diallel Crosses

When, in the course of breeding for resistance to lymphomatosis at Cornell University, it became evident that the chief limiting factor was a scarcity of proven sires, a change in the system of breeding was adopted which quickly overcame that difficulty. It is called the "double shift," except when there are three shifts, in which case it is called the "triple shift."

The first of these terms means merely that, in all the pens available for cockerel testing, one series of males is used in the first half of the season and is then replaced by an entirely different lot for the last half of the season. In the first year of its use, the double shift proved so satisfactory that it was abandoned and replaced by the triple shift, which is still better. By its use, 30 cockerels are tested annually in 10 breeding pens, and there is no dearth of proven sires.

Transfers of the males are arranged to accommodate weekly settings of eggs in the incubators on Tuesday. To illustrate, shift I goes out of the breeding pens on Friday afternoon and is replaced on Sunday in the late afternoon. Eggs laid to Monday night are credited to shift I, but eggs from Tuesday to Saturday, when paternity is doubtful, are marketed. Beginning on

the following Sunday, eggs are saved for incubation and credited to shift II, this being 7 days after the introduction of those
cockerels and 9 days after the removal of shift I. In *Genetics of
the Fowl, Chapter* 13, evidence is cited to justify the assumption that the influence of the replacing male supplants that of
the replaced male within 9 days. The details of working out a
program of triple shifts are perhaps best understood by study
of the actual records for a typical case, as presented in Table 5.

Table 5. Details of the Operation of Triple Shifts for Testing Cockerels at
Cornell University in 1947

Shift	Males intro-duced, date	Period of saving eggs		Days of lost eggs	Males re-moved, date	Daughters per male at 6 weeks	
		Dates	Days			Average	Range
I	Jan. 3	Jan. 29-Feb. 17	20	None	Feb. 14	39	26-49
II	Feb. 16	Feb. 23-Mar. 17	23	5	Mar. 14	58	33-90
III	Mar. 16	Mar. 23-Apr. 14	23	5	Immaterial	50	16-81

These data pertain to a hatching season of 10 weekly settings, beginning on Feb. 11, of which, because of the two intervals between shifts, only 8 included eggs from the cockerel-
testing pens. Those were 2 settings from the first shift and 3
from each of the others. The breeding pens contained 16
females each. The comparatively small number of daughters
per sire in shift I resulted from somewhat low fertility in the
early settings and also because the period of saving eggs was 3
days shorter *(i.e.,* 48 hen-days) than for the other two shifts.
Troubles with males of low fertility are inevitable but are less
serious in a shift of 3 to 4 weeks than when one such male is
used all season.

It will be noted that altogether 10 days' eggs were lost (marketed) because of the intervals between shifts. Anyone in a position to practice artificial insemination should note the evidence of Warren and Gish (page 450) that when that procedure is used little error in paternity is entailed if every egg is saved and none discarded. Most of those on the second day after use of the second male can safely be assigned to him. With such a program, and to get such results, it would be necessary to have the replacing males well trained prior to the date on which they are to be used.

While the primary purpose of the double and triple shifts is to provide progeny tests for a large number of cockerels in one season, another advantage lies in the fact that they permit better tests of the genotypes of the dams. If a hen produces superior offspring by one male, that result might be attributable to (1) her own good genes, or (2) his good genes, or (3) complementary action of genes from both. The last of these three possibilities is what the breeder calls "nicking." Individuals that nick with one mate might produce inferior offspring by another. It is desirable to find the hens with genotypes such that they will produce superior offspring with any male or with most of them. Obviously, such birds are more likely to be revealed by matings in turn with the two or three males that are used in the double or triple shifts than when mated all season with a single male.

In these cases, the value of the female is not determined merely by the performance of her own offspring by two or three sires. Each of the sires is first evaluated according to the average performance of his offspring from all the hens in the breeding pen. Each dam is then appraised by the record of her daughters by each sire in relation to the averages for each sire. Such an evaluation shows whether or not she excels the other hens in her breeding pen.

This method of determining a genotype with respect to multifactorial characters was devised by the distinguished Danish ichthyologist, Johannes Schmidt (1922), who designated it as diallel crossing. It should interest poultry breeders to know how he came to demonstrate the value of diallel crosses in poultry breeding. Schmidt will be best remembered for his studies with eels. In the course of tracing North American eels to rivers in that continent and European eels to rivers in Europe, both from their common breeding grounds in the western Atlantic, he became interested in differences in the number of vertebrae by which these two types are distinguished. This interest was extended to other species, and Schmidt was able to prove experimentally in trout and in the fowl that differences in the genotype with respect to number of vertebrae could be accurately detected by diallel crosses. The procedure is equally valuable for any other multifactorial character.

Importance of the Environment

Most inherited characters of economic importance in the fowl can be modified by the environment. It is essential to progeny testing that differences caused by genes should not be obscured by variations induced by the environment. Clearly, the environment should be as uniform as possible for all birds under test, but it is easier to state that policy than to maintain it. It is easy to mix representatives of all families at random, to give all birds the same feed, to house them in flocks of comparable size, and to do all the other obvious things. It is not so easy to keep the late-hatched chicks from getting coccidia, to which their early-hatched sisters are better able to develop resistance. It is most convenient to vaccinate all birds for chicken pox or for Newcastle disease at one time, but some of them are much closer to laying than others when this is done. In addition to such complications there are others that may be entirely unsuspected.

A good example of unsuspected environmental modification was encountered by the author and his associates in breeding strains of White Leghorns resistant or susceptible to lymphomatosis. It was the custom each year to start the first hatch in one large brooder house, F, to start the second in another similar house, B, and thereafter to alternate the remaining eight weekly hatches in this way between the two houses. At 2 weeks of age, all chicks went to the rearing range, thus clearing the brooder houses for the next hatch of chicks. Not until this had gone on for 7 years was it discovered that chicks started in house F were annually exposed more severely to lymphomatosis than those started in house B. Moreover, that brief difference, though not evident at 2 weeks of age, was reflected in the mortality after the pullets became mature (Hutt *et al.*, 1944). Each year, as a result, pullets of the odd-numbered hatches (first, third, fifth, etc.) experienced higher losses from that disease than did their sisters of the even-numbered hatches. The 7-year averages for two strains shown in Fig. 9 reveal the remarkable influence of this brief difference in the environment. Because at that time there were no double or triple shifts, representatives of all families shared equally in the two kinds of exposure; hence familial differences were quite evident, although less so than they would have been had all the chicks been severely exposed. Once the difference was detected, all chicks were started in house F; all thus received the severe exposure, and selection became immediately more effective.

The extent to which such a difference in the environment can affect the appraisals of males being tested for viability of their offspring is illustrated in Table 6 by the records for 3 of 17 males so tested in one year. For each of those 17, one half the chicks were severely exposed (in house F); the other half (alternate hatches) went directly to the rearing range and hence were exposed only lightly.

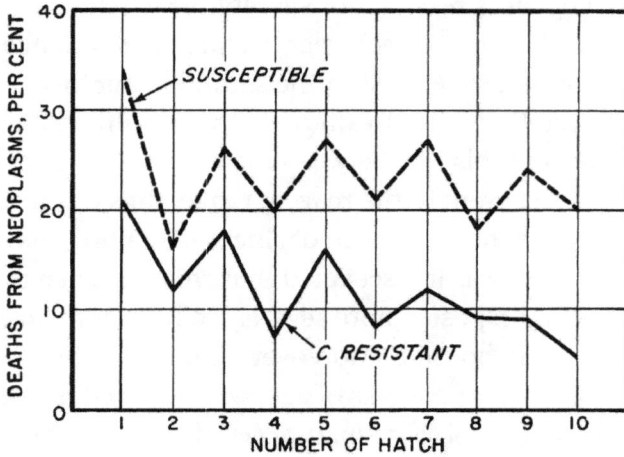

FIG. 9. *Deaths from neoplasms in two strains of Leghorns, one resistant to lymphomatosis and the other susceptible, in birds of 10 weekly hatches between Mar. 1 and May 10. Each point is the unweighted average of 7 years' data. The graphs show the effect of the more severe exposure of the odd-numbered hatches.* (From Hutt *et al.* in *Poultry Sci.*)

Daughters tested under one environment ranked ♂ A best among 17; under the other environment he ranked next to the last. A similar reversal is shown for ♂ C. Clearly the breeding value of an individual, when measured by the progeny test, must depend greatly upon the environmental conditions under which that test is conducted.

Table 6. Influence of the Environment on the Ranking of 3 Males among 17 Tested for Viability of Their Offspring

Male	Lightly exposed daughters			Severely exposed daughters		
	Housed at 160 days, number	Died by 500 days, per cent	Rank among 17	Housed at 160 days, number	Died by 500 days, per cent	Rank among 17
A	33	12	1	38	58	16
B	73	16	2	82	39	7
C	36	19	6	41	61	17
Average for all 17		24.7	8.5		43.6	8.5

One important environmental influence that must always be considered is the length of day or the amount of available light. The relation between variations in that factor and differences in age at first egg, in persistency, and in egg production are discussed at length in *Genetics of the Fowl, Chapter* 10. It will suffice here to remind the reader that its influence makes it impossible to compare late-hatched chickens and early ones with respect to these three variables unless suitable corrections are first made in the records. This is particularly important when double or triple shifts of cockerels are tested. That direct comparisons between sires in these different series are impossible is illustrated by the tests in 1 year for 15 cockerels of one strain, there being 5 of these in each shift. The females in the five pens were constant through the entire breeding season. For daughters of these males, three averages considered in December were as follows:

	First Shift	Second Shift	Third Shift
Age at first egg, days	178	188	196
Eggs to Dec. 1, number	55	40	16
Mortality, 6 weeks to Dec. 1	2.3%	6.9%	5.8%

In addition to the expected differences in age at first egg and the inevitable one in egg production, it seems probable that the early-hatched chicks from the first shift had escaped some influences that caused higher mortality in their later hatched half sisters. Corrections could be made with respect to age at first egg, but not very satisfactorily in the other measures of desirability. Such corrections are unnecessary if comparisons are first made only among the five contemporary males of any one series. After that, one can estimate for the whole 15 males which ones excel in greatest degree the averages for their respective groups.

The Use of Proven Sires

Since this section on progeny testing has dealt mostly with the necessity of using proven sires and with methods for finding them, it is appropriate to include a few words about the use of such birds after they have been discovered. Although this is the most important part of the poultry breeder's program, it requires little comment. Obviously, one uses a superior sire to beget the greatest possible number of offspring. Moreover, such sires should be used as long as they will produce fertile eggs or until better ones are discovered by the progeny tests of later generations. Even if the fertility of a good old sire drops to 40 per cent in his fifth breeding season, the 30 offspring that he may then have are more valuable than 200 secured from some mediocre cockerel being tested for the first time. This is particularly so because the wise breeder will have mated the best proven sire to some of the best proven dams in the hope of thus preserving the desirable genes of both sides for transmission to future generations by their sons.

While 3 weeks' hatching eggs may suffice to provide a progeny test for a cockerel, the good proven sire should be used continuously throughout a breeding season as long as is compatible with the limitations of latitude and of good management. He should be mated with many hens. Sometimes such a male seems to maintain better fertility if his mates, to the number of 20 or 30, are divided in two pens, with the sire switched on alternate days from one pen to the other. A still larger number of hens might be used if artificial insemination be practiced.

The proportions of the available breeding pens that should be devoted to cockerel testing and to proven sires will have to be decided annually by the breeder. They will vary according to his needs and the number of proven sires available. When these last are few, it may be desirable to devote 80 per cent of the available space to cockerels. Ten pens would thus permit testing 24 cockerels in three shifts and leave two pens for the use of the best proven sires throughout the season. Conversely, while few breeders are likely to be embarrassed by a surplus of superior proven sires, if that happy situation should ever arise it might be feasible to omit cockerel testing for one season.

Selection of Females

The emphasis here placed upon proven sires and the comparatively slight discussion of proving dams should not be interpreted as any indication that the genes from the dam are unimportant. Both sexes contribute equally to the inheritance, except for such sex-linked genes as females may receive from their sires. However, since a single sire may transmit his good or bad genes to the 130 offspring of 16 different dams, whereas any one of those dams is responsible for only 8 or 9 of that number, the old adage that "the sire is more than half the flock" should never be forgotten by the poultry breeder.

For the cockerel tests it is desirable to use as good females as are available in the numbers needed. They may be the next to the best hens left after the best ones have been allocated to the proven sires, or they may be pullets. The important thing is that each pen in the cockerel-testing program should have a random assortment of the available females. This is not hard to arrange when there are 16 birds, or thereabouts, per pen. Exceptions can be made as needed. For example, if one family from which cockerels are tested has all desired good points except body size, three brothers of that family may be used (in successive triple shifts) in the same breeding pen, and it can be made up to include females above average in size. Any such adjustment in January must not be forgotten in the following December when all the cockerels are being compared.

In addition to their main purpose, or perhaps as a part of it, the cockerel-testing pens will bring to light each year many females of superior merit. By utilizing annually the best of these, there should be no difficulty in maintaining at full strength the company of proven dams reserved for the breeding pens of the proven sires.

Inbreeding

Most poultry breeders become concerned about inbreeding as soon as their stock has been developed to the point at which they can view it with some pride of achievement. They would then like to maintain its good points, to improve it still further, perhaps with respect only to some minor character, but at all costs not to lose what they have gained by careful breeding. At that stage their problem is often whether to risk the dangers of inbreeding by continuing to use their own males or the other danger of bringing in undesirable inheritance if new blood be introduced from some other flock.

In the past many breeders of livestock have considered inbreeding as a word to be spoken only in whispers—its effects as abhorrent as the plague. Poultrymen, who have really less to worry about in this respect than have breeders of larger animals, have (in the past) been particularly prone to consider that anyone's temporary hard luck in the matter of fertility, hatchability, egg size, or body size has resulted from the stock's becoming too closely inbred. It now seems probable that this attitude may be abruptly changed and that, for some breeders at least, inbreeding may become an objective just as desirable as high records of egg production have been for years. This is happening merely because poultrymen know that suitable inbred strains of corn will give phenomenal yields when crossed. The same thing is now being demonstrated with poultry and with swine. For this reason, the future will probably see some poultrymen as anxious to prove that they have an inbred strain as they were to deny it in the past. It is to be hoped that the effort to be in the swim will not tempt any inexperienced swimmer to dive in waters beyond his depth.

Rates and Measures

The general effect of inbreeding is to increase the number of homozygous pairs of genes and to decrease the heterozygosity. It thus makes the inbred populations more uniform and differentiates families or lines. The unfortunate consequences that have caused its disrepute result from the inevitable lethal genes and others with undesirable effects. When such genes are made homozygous, death or reduced viability follows.

The closest kind of inbreeding possible in higher animals is the mating of brothers with sisters. With that system one does not attain as much homozygosity in 17 generations as can be brought about by 6 generations of self-fertilization in maize, a point worth remembering by anyone hoping to produce hybrid poultry that is as profitable as hybrid corn. On a theoretical basis, it would take over 20 generations of brother X sister matings to yield individuals that resemble each other as much as the two members of a pair of identical twins. In practice, even that is not likely to be attained, because of new mutations. Moreover, attempts thus far to get highly inbred strains of fowls by this most rapid method have been so unsuccessful as to suggest that a somewhat less intense degree of inbreeding may be preferable.

One such milder form of inbreeding is the mating of a male with his half sisters. Wright (1931) calculated that, whereas matings of full siblings reduce the residual heterozygosity by about 19 per cent in each generation, matings of a male with his half sisters reduce it only 11 per cent in each generation. Obviously, this is somewhat safer.

Contrary to common opinion, continued breeding within one flock does not result in rapid inbreeding, so long as the breeding stock is chosen at random and several different sires are used each year. When this is done, according to Wright, the heterozygosity is reduced in each generation by approximately 1/8N, where N is the number of males used, provided that the number of females greatly exceeds the number of males. This formula has a special significance for poultry breeders because it is applicable to exactly the conditions that prevail in most sizable flocks to which new blood is not introduced. If 3 males be used per season, the remaining heterozygosity is reduced only about 4.2 per cent each year; if 10 be used, the figure is about 1.2 per cent. Such a slow approach to homozygosity is not likely to cause trouble in any flock. It is a system that has been successfully used in many.

It is customary to measure the amount of inbreeding in any individual by a formula devised for this purpose by Wright (1923). When X represents the inbred individual and A some ancestor common to X's sire and dam, the coefficient of inbreeding F_x is determined by the formula

$$F_x = \sum [0.5^{n+n'+1}(1 + F_A)]$$

where n is the number of generations from the sire of X back to the common ancestor, n' is the same for the dam of X, and F_A is the coefficient of inbreeding of the common ancestor A. If A be not inbred, that portion of the formula can be ignored, and if A be further back than four generations, its contribution is relatively unimportant. The sign Σ means merely that one must determine separately each different link in the pedigree by which the sire and dam are related and then add them up. The coefficient of inbreeding is a measure of the degree of relationship between the sire and dam, corrected, when necessary, for the inbreeding of a common ancestor A.

By way of illustration, and as a convenience to those who may wish to measure inbreeding, the component parts of the coefficient of inbreeding that are most frequently used are shown in Table 7. It is evident from these that a common ancestor more than four generations back from the sire or dam of the individual under consideration can affect the coefficient of inbreeding very little.

Table 7. Contributions of Some Common Relationships To the Coefficient of Inbreeding

Number of generations to the common ancestor		Contribution to the co-efficient of inbreeding
Behind one parent	Behind the other parent	
0	1	$(\frac{1}{2})^2 = 0.25$
1	1	$(\frac{1}{2})^3 = 0.125$
1	2	$(\frac{1}{2})^4 = 0.0625$
2	2	$(\frac{1}{2})^5 = 0.0312$
2	3	$(\frac{1}{2})^6 = 0.0156$
3	3	$(\frac{1}{2})^7 = 0.0078$
3	4	$(\frac{1}{2})^8 = 0.0039$
4	4	$(\frac{1}{2})^9 = 0.0019$

Some pedigrees may show only one of these relationships. With those which have more than one, the contributions of each relationship must be added. To determine the amount of inbreeding in a flock, strain, or family, it is necessary to determine the average for a representative sample. A coefficient of 0.25 is said to show 25 per cent inbreeding.

A single brother x sister mating produces a chicken (or calf, or anything else) for which the coefficient is 25 per cent. With two and three generations of the same system, it becomes 37.5 and 50 per cent, respectively. For offspring from one mating of a male with his half-sisters, the coefficient is 12.5 per cent. It was shown earlier that, according to one of Wright's formulae, the degree of inbreeding in a flock is slight when no new blood is added over several years, so long as several males are used each year and inbreeding is not deliberately sought. For two such strains at Cornell, in which no new blood was introduced during 11 years, the coefficients of inbreeding were found to be only 7 and 8 per cent. In a third strain, to which new blood had been added, inbreeding happened to be 13 per cent.

Line breeding refers to the continued mating back to descendants of some particular ancestor in order to increase the contribution of that animal to the inheritance. It is inbreeding, but of a degree less intense than that resulting from the mating of closer relatives.

Effects of Inbreeding in the Fowl

The effect of inbreeding that is most obvious, perhaps because it is soonest evident, is a marked decline in hatchability of the eggs. In one of the first attempts to produce highly inbred fowls, Cole and Halpin (1916) found that in three generations of brother x sister matings the hatchability of fertile eggs, which was originally 67 per cent, declined to 49, 41, and 18 per cent. In this experiment, hatchability was not considered in the selection of breeders, but in a later similar trial, in which such selection was practiced in the hope of maintaining the stock, hatchability again declined so much that the inbred line could not be maintained (Cole and Halpin, 1922). Similar declines in hatchability, in varying degrees and resulting from different kinds of inbreeding, have been reported by Dunn (1923, 1928), Jull (1929, 1929a, 1933), Dumon (1930), Dunkerly (1930), Dudley (1934), and others.

The uniformity with which hatchability has been reduced in the different stocks studied by these investigators confirms the assumption that many lethal genes are widespread in *Gallus gallus* and assures the young graduating geneticist that there are still fields for him to conquer. For those who cannot forget the fields of waving hybrid corn, some encouragement is seen in the less frequent reports of those investigators who have been able to maintain a fair level of hatchability even in stock that is rather highly inbred. Waters and Lambert (1936, 1936a) developed a high degree of homozygosity in White Leghorns by degrees of inbreeding less intense than that of brother x sister matings and maintained careful selection for hatchability. As a result, the average hatchability was still well above 60 per cent when the coefficients of inbreeding ranged from 41 to 82 per cent. With other stock, Waters (1945) was able to raise four generations from brother x sister matings without any great decline in hatchability. Similarly Dumon (1938) developed by rigorous selection some inbred lines in which hatchability was even superior to that in birds not inbred at all.

From all these reports one must conclude that, while the closest inbreeding is likely to cause the extinction of most lines, if enough of them are started and if rigorous selection be practiced, some of them will prove capable of adequate repro- duction. The prospects for getting highly inbred lines with the least losses would seem to be best with inbreeding less intense than the continued mating of full brothers and sisters. Since this is a slower procedure, those preferring speed to safety will gamble on the more direct approach. They should be prepared to utilize a proportionately greater number of birds and also of bank notes.

Reports of the effects of inbreeding on egg production, fer- tility, body size, and other economic characters vary consider- ably. That is not surprising. There is little point in attempting to reconcile the irrefutable evidence of Dunn (1923), Jull (1933), Hays (1934), and others that inbreeding lowers egg production and raises age at first egg with the equally irrefutable evidence of Waters and Lambert (1936) that it does not do so in every case. It is to be expected that inbreeding will differentiate families, although tending to increase uniformity within families. Accordingly there should be, not only differ- ences in the strains inbred by different investigators, but also equally great differences among the inbred families of any one of them.

While the differentiation of inbred families has been noticed by most investigators, the expected increase in uniformity is not always so evident. Dunn (1928a) found that eight inbred families of Leghorns differed in the length and proportions of certain bones of the limb and also in the shape of the cranium. Variability in these skeletons was slightly less than in outbred stock. In a flock brought in 9 years to an average coefficient of inbreeding of 60 per cent, Shoffner (1948) could see no consistent corresponding decline in variability. His inbred, later generations were more uniform than the earlier ones in egg production, age at first egg, and hatchability, but not in body weight or in size of egg.

Prepotent Sires

One of the objects of inbreeding is to get individuals that are homozygous with respect to desirable genes manifested by relatives. In such cases, inbreeding is risked in order to concentrate the blood (genes) of these outstanding relatives. The breeder hopes to get thus some sires that will pass desirable traits to a majority of their offspring.

Such a sire is said by the breeder to be prepotent and by the geneticist to be relatively homozygous. To get one without too great risk, it is sometimes good practice to mate a good proven sire or a promising cockerel (1) with some close relatives, such as sisters, daughters, or dam; (2) with more distant relatives; and (3) with unrelated females. Even if the first of these combinations should prove disastrous, some good genes may be salvaged by the others.

The Utilization of Hybrid Vigor

Another purpose for inbreeding is to develop highly inbred lines suitable for crossing with the hope of getting from such crosses enough hybrid vigor to cause greater productivity or viability (or both) than is already available in stock not inbred. As this type of inbreeding may be tried by quite a few breeders in the future, it is desirable to consider what happens when inbred lines are crossed. They may be utilized in three different ways, each of which is here considered separately.

Crosses of Inbred Lines of One Breed

In evaluating the performance by progeny of crosses between inbred birds, one must not conclude that merely because such progeny excel the inbred parents some special merit of such crosses has been demonstrated. From what is known of similar crosses in other species, it is to be expected that highly inbred birds will usually produce (when crossed) offspring superior to both parents The important question is whether or not the hybrids will excel improved stock that has not been deliberately inbred.

The data of Dunn (1928) and Jull (1930, 1933), showing that hatchability is improved when inbred strains are crossed, provide no evidence that it improved to the point of excelling stock not inbred. In fact, crossings by Jull (1933) yielded hatches 12 to 37 per cent lower than those of the original outbred stock. In four groups of his pullets from such crosses, the mean egg production was 35 to 72 eggs lower than in their outbred ancestors. Maw (1942) found that in some crosses between inbred lines of Iowa Leghorns the hybrids did excel controls not inbred, but since those controls laid only 73 eggs in 20 weeks, the significance of the difference might be questioned. Similarly, the data of Pease (as given by Maw) show that hybrids from two different strains of inbred Leghorns (one bred in Iowa and one at Reaseheath in England) excelled both the parental strains in egg production but barely equaled the record of Leghorns not inbred.

From these results it w°uld seem that any advantage that might be gained merely by crossing two inbred lines of one breed has yet to be demonstrated. It is to be expected that some of these might yield better results than have been reported to date.

Double Crosses of Four Inbred Lines

The negative results reviewed in the previous paragraphs have little or no bearing on the value of inbred lines for crossing. This is merely because anyone seeking hybrid vigor from the use of inbred strains is better advised to cross such strains from different breeds than to cross them within one breed. Over a period of several years, Knox (1946) found that females from inter-strain crosses of inbred Rhode Island Reds had an average production of 198 eggs, which differed little from that of similar females not inbred. However, when crosses were made between inbred Reds and inbred Leghorns, the mean production was 224 eggs. It is to be expected, from what is known about such crosses in maize, that the maximum benefits from hybrid vigor would result when the progeny from two inbred strains, *A* and *B*, are crossed with the progeny from two other inbred strains, *C* and *D*, to produce what is commonly called a "double-crossed hybrid." Experimental evidence of the value of such procedures in the fowl is still almost negligible. Dumon (1930) provided some evidence that hatchability is increased, and Maw (1942) cites a double cross in which mortality of chicks was apparently greatly reduced but egg production was nothing unusual.

Meanwhile, the production of double-crossed hybrid chickens on a commercial scale has begun and gives every promise of expanding as rapidly as inbred strains can be developed. The performance of such stock as is now available is good enough to create a tremendous demand for it, and this indicates that the double-crossed hybrid fowls have an assured future. Extensive studies are desirable to find out what kinds of inbred strains and what degrees of inbreeding are most suitable for such crosses. Because it is more difficult to get high homozygosity in the fowl than in maize, the poultry industry should not expect too much too soon from the hybrids.

Top-crosses on Outbred Stock

It seems probable that one of the most important benefits from highly inbred lines may be obtained by top-crossing the inbred males on flocks that are not inbred. The significant evidence from 5 years of such crosses presented by Waters (1938) shows an increase in hatchability and a decrease in mortality that are remarkable (Table 8).

Table 8. Effect of Top-crossing on Some Economic Characters in Leghorns; Averages for 5 Years (Walters, 1938).

	Fertility	Hatchability of fertile eggs	Mortality	
			To 8 weeks	To 24 weeks
1. Leghorns, not inbred	75.9%	80.8%	8.1%	15.8%
2. Leghorns, inbred	83.6%	72.1%	14.2%	23.0%
3. Top-crosses; ♂♂ of 2 x ♀♀ of 1	85.7%	85.8%	5.8%	10.3%

Evidently the Leghorns that were not inbred were unusually low in fertility. At any rate, the apparent superiority of the other two groups in that respect is not significant. In hatchability, the top-cross progeny were consistently superior in every year regardless of which of the seven inbred families provided the males for crossing. (This refers to hatchability of the eggs laid by the random-bred Leghorns, whence came the top-cross progeny.) Mortality to 24 weeks was significantly lower for the top-crosses than for the Leghorns not inbred. To these important findings should be added the confirmatory evidence of Maw (1942) that top-crossing produces birds characterized by low mortality and comparatively high production, but it is not clear that there were any significant differences between his top-crosses and the random-bred controls. Results to be expected from such top-crosses will doubtless vary according to the genotype and the degree of homozygosity of the inbred strain used. The results of Waters suggest that the possibilities of top-crossing merit considerable investigation.

Crosses of Strains not Inbred

Apart from the use of strains deliberately inbred for crossing, a certain amount of hybrid vigor is to be expected in some crosses between flocks, strains, or breeds that may be considered entirely free from inbreeding. In the broader interpretation of that term, some inbreeding is present in any of those three commonly cited divisions of the poultry population.

A flock of White Leghorns in which inbreeding has been avoided will still be homozygous for certain genes affecting color of the plumage, ear lobes, skin, and eggshells and for others affecting the comb or other characteristics. To the extent that these genes are included, while those which distinguish Wyandottes, Houdans, or other breeds are excluded, the White Leghorns, as a breed and variety, are somewhat inbred. The same applies to any other breed. Furthermore, within any color variety of one breed *(e.g.,* White Leghorns), strains may be differentiated by selection for different objectives on the part of different poultry breeders and by natural selection in different environments. To the extent that uniformity is thus increased within the strain, so is the inbreeding.

The scale in Fig. 10 illustrates the fact that there is really a steady gradation in the different degrees of inbreeding and no sharp boundary between any of them. It is as difficult to specify any point on this scale at which inbreeding begins or stops as it is to mark a spot on the thermometer to divide hot from cold. One might expect, therefore, that some measure of hybrid vigor should result from crossing strains, varieties, or breeds. Other things being equal, it should be proportional to the number of pairs of genes by which the crossed groups differ. That difference may be, in some crosses, so slight as to cause no perceptible hybrid vigor, while in others it may be quite pronounced.

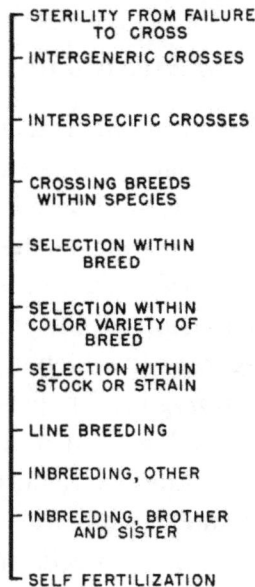

STERILITY FROM FAILURE
TO CROSS

INTERGENERIC CROSSES

INTERSPECIFIC CROSSES

CROSSING BREEDS
WITHIN SPECIES

SELECTION WITHIN
BREED

SELECTION WITHIN
COLOR VARIETY OF
BREED

SELECTION WITHIN
STOCK OR STRAIN

LINE BREEDING

INBREEDING, OTHER

INBREEDING, BROTHER
AND SISTER

SELF FERTILIZATION

FIG. 10. A scale showing gradations in the degree of inbreeding between the two extremes.

Data on the value of various crosses of strains within a breed are scarce, but this lack may be overcome by the numerous interstrain crosses now being made as a result of the recently stimulated interest in the utilization of hybrids of various kinds. Warren (1942) found some hybrid vigor from crosses of two American strains of White Leghorns but less than from crosses of distinct breeds. Older poultrymen will recall some phenomenal records made by a California breeder who crossed American and Australian White Leghorns.

Crosses Between Two Breeds

Obviously more hybrid vigor is to be expected from crosses of different breeds than from crosses of strains within a breed merely because breeds differ genetically more than strains do. Reports from different investigators show that, when two breeds are crossed, any stimulation resulting may be conspicuous with respect to some items but entirely lacking so far as others are concerned. There is general agreement that hatchability (of the eggs from which the cross-breds hatch) is raised by 5 to 20 per cent in most crosses (Warren, 1927, 1942; Byerly et al, 1934; Knox and Olsen, 1938; and others). A reminder that neither this result nor any other manifestation of hybrid vigor is to be expected without fail in every cross is found in the data of Dudley (1944) showing no significant increase in hatchability for crosses of White Leghorns and Rhode Island Reds.

Some of the best evidence on this score comes from the few cases in which the same hens have been mated with (1) males of their own breed and (2) males of another (Table 72).

Faster growth of hybrid chicks than of purebreds has been noted by most investigators (Warren, 1927, 1942; Bice and Tower, 1939; Knox et al., 1943). The same thing has been observed in turkeys by Asmundson (1942). As is to be expected, there are exceptions to this rule. Among 11 interbred crosses of various kinds made by Horlacher *et al.* (1941) to determine the suitability of some of them for the production of broilers, the hybrids at 10 weeks weighed in three cases significantly less than birds of the heavier parent breed. In four crosses, the hybrids did not differ much from the heavier parents, but in four others they were significantly heavier. The cross White Wyandotte ♂ x Rhode Island Red ♀ proved consistently good for 3 years. At 12 weeks of age broilers from this cross weighed 11 per cent more (125 gm.) than the pure Wyandottes, which in turn were 134 gm. heavier than the Rhode Island Reds. Unfortunately, these hybrids apparently sold at a discount because of the Columbian pattern in their plumage.

The hybrid vigor resulting from crosses between suitable breeds is of great value to those raising broilers, but one must not expect the accelerated rate of growth to be maintained much beyond 12 weeks, if any. At maturity, birds of the F1 generation from parents differing in size are more likely to be intermediate in size than to exceed either parent.

An important point making the hybrid vigor of crossbred broilers of special value is that their more rapid growth does not require proportionately more feed. Evidence that the hybrids utilize their feed more efficiently than purebreds during the period of rapid growth has accumulated from Bice and Tower (1939), Hess *et al.* (1941), Horlacher *et al.* (1941), and Mehrhof *et al.* (1945).

Table 10. Hatchability of Eggs Laid by Hens Mated with Males of Their Own Breed and with Others

Breeds crossed	Hatchability, for hens mated	
	With males of same breed	With males of other breed as shown
Warren (1942):		
White Leghorn ♀ ♀ A x Barred Plymouth Rock ♂ ♂	61%	80%
White Leghorn ♀ ♀ B x Barred Plymouth Rock ♂ ♂	84%	86%
White Leghorn ♀ ♀ C x Barred Plymouth Rock ♂ ♂	71%	80%
Funk (1934):		
Barred Rock ♀ ♀ A x Rhode Island Red ♂ ♂	57%	74%
White Leghorn ♀ ♀ C x White Plymouth Rock ♂ ♂	71%	80%

The superior viability of hybrid chicks has been pointed out by several investigators (Warren, 1927, 1942; Horlacher *et al.* 1941; Dudley, 1944). In Horlacher's broilers, the hybrids from 8 different crosses out of 11 suffered less than 10 per cent mortality to 12 weeks of age, but only 2 of 7 lots of purebred birds did as well. Concerning viability of adult hybrids, there is less evidence, and it is not conclusive. Warren (1942) could find no consistent indication that hybrid vigor is reflected in lower mortality of adults, nor were the data of Dudley (1944) conclusive. On the other hand, Knox *et al.* (1943) found the viability of crossbreds after 20 weeks to be better than that of purebreds in 9 out of 10 comparisons. The differences were 6 and 10 per cent in 2 different crosses.

With respect to the egg production of hybrids, it would appear from such data as are available that, if it is increased at all above the level of the two parent breeds, the differences are slight and vary greatly depending on the strains used. Warren (1942) found that in most of such crosses the hybrids laid as well as the higher producing parent breed, or better. Some crosses seemed to yield better results than others. Zorn and Krallinger (1934) crossed Leghorns and Faverolles and found the F1 generation to be only intermediate between the two parent breeds in laying ability. Knox *et al.* (1943) made a number of crosses among Rhode Island Reds, White Wyandottes, and Light Sussex but found none of them to surpass the pure Rhode Island Reds in egg production. The hybrids of Dudley (1944) from White Leghorns and Rhode Island Reds had averages of 197 and 204 eggs, against 194 for pure Leghorns and 197 for pure Rhode Island Reds. It would seem that these records of the hybrids might have looked still better if 20 to 25 per cent of the pure Rhode Island Reds had not been culled for late maturity.

From a review of these and other experiments, one is led to conclude that the influence on egg production of hybrid vigor from simple crosses between breeds might perhaps be better revealed by experiments in which the same males and females are mated (in diallel crosses) with birds of their own breed and of others.

Crosses among Three Breeds

Advocates of crossbreeding have usually attempted to soothe the protesting devotees of the pure-breeding cult with the assurance that the F1 hybrids should never be used for breeding. Nevertheless, swine breeders have found that, when such F1 animals are crossed with some third breed, the offspring are even better than the parents. If such an accumulation of hybrid vigor can be made in the fowl, it has yet to be demonstrated. Neither Warren (1942) nor Knox *et al.* (1943) could find any marked superiority of the progeny from three breeds over that from two breeds. It is quite possible that other breeds or strains thus mixed might yield different results.

Interspecies Hybrids

Crosses between related mammalian species are commonly made to produce two very useful domestic animals. One of these is the mule, a familiar example of hybrid vigor. Crossing the zebu (Brahman) cattle with the species more familiar to most readers of this book is sometimes followed by further backcrosses to secure a combination of the productivity of some breeds with the zebu's ability to tolerate heat and to withstand tick-borne diseases. In fowls, no such useful hybrids between species have yet been developed.

The Ceylonese Jungle Fowl, *Gallus lafayettii*, crosses readily with the domestic fowl, according to experiments of the Ceylon Poultry Club reviewed by Lotsy and Kuiper (1924) and others carried out by Ghigi (1934). The hybrids are fertile among themselves and in backcrosses to the domestic fowl.

The Grey Jungle Fowl, *G. sonneratii,* also produces fertile hybrids with the domestic species (Lotsy and Kuiper, 1924; Ghigi, 1922). These are apparently fertile in both sexes and will breed *inter se* and in back-crosses.

The Black Jungle Fowl, *G. varius,* apparently does not cross so readily, but Houwink (1921) reported that he had raised several hybrids between that species and *G. bankiva (galius),* some fertile and some sterile. Ghigi (1934) found fertility of this hybrid to be limited in the male. His one F1 female laid eggs, but none were fertile.

Intergeneric Hybrids

To complete the record of all possible types of mating between the limits of self-fertilization and complete incompatibility shown in Fig. 10, a brief reference to crosses of the fowl with birds of other genera is in order.

With Pheasants

The most common hybrids of the domestic fowl are those with different geographic races of ring-necked pheasants, which have been designated by ornithologists as *Phasianus colchicus colchicus, P. c. torquatus,* and *P. c. mongolians.* A number of these were collected by Poll (1912), who reviewed previous records of such hybrids and showed that their invariable sterility results from incomplete gameto-genesis in both males and females. In a number of hybrid males studied by Cutler (1918), spermatogenesis went no further than the formation of primary spermatocytes.

Sandnes and Landauer (1938), noting that with a single exception all the recorded hybrids of this type had come from the cross of fowl ♀ x pheasant ♂ and that females were rare among the hybrids, made the reciprocal cross by artificial insemination. Embryonic mortality was 43 per cent, but 58 embryos died late enough to permit identification of sex, and 97 chicks were hatched. Of these, 20 were still alive (and not sexed) some months later. Excluding these last, the sex ratio in this cross and that for recorded hybrids of the reciprocal cross were as follows:

	Hybrid offspring	
	Males	Females
Pheasant ♀ x fowl ♂	66	69
Fowl ♀ x pheasant ♂	23	6

The deficiency of females in the second of these reciprocal crosses conforms to expectation according to Haldane's rule. This holds that, when, in the offspring of an interspecific cross, one sex is missing, rare, or sterile, that sex is the heterogametic one.

The transmission of mutant characters in such crosses was studied by Serebrovsky (1929) in Poll's hybrids. The general rule seemed to be that mutations dominant in the fowl were also dominant in the hybrids, examples being bare neck, extension of black, foot feathering, and dominant white. Sex-linked barring was transmitted by one hen to her hybrid sons but not to her daughter. The manifestation by such hybrids of dominant white from a Leghorn parent has also been reported by Danforth and Sandnes (1939).

With the Guinea Hen

From the classification of the Galliformes given in *Genetics of the Fowl*, Chapter 1, it can be seen that the guinea hen, *Numida meleagris,* belongs to an entirely different family from that of the pheasants. Although few offspring from crosses of such widely different species are found in the animal kingdom, a number of hybrids between the domestic fowl and the guinea hen have been recorded. Some of the earlier cases were reviewed by Guyer (1909, 1912), who studied five specimens 3 years old, all males. A hybrid from a guinea ♀ x Rhode Island Red ♂ was described by Viljoen and de Bruin (1935). The four hybrids reported by Funk (1938) from pearl guinea hens and a Rhode Island Red father are particularly interesting because, contrary to the usual predominance of males, some of them were females that even laid eggs. In all these hybrids, the comb is absent or rudimentary.

Attempts to produce this hybrid by artificial insemination have not been successful. Marchlewski (1937) and Owen (1941) found the fertility of eggs to be less than 15 per cent. Most of the embryos died early, but Marchlewski had two chicks that died at 26 days of incubation after starting to hatch.

With the Peafowl

Two adult hybrids from a peacock, *Pavo cristatus,* and a Buff Cochin hen were described by Trouessart (1920), but attempts of Tiniakov (1934) to produce such a hybrid by artificial insemination were only partially successful. From five fertile eggs, one chick hatched (after 25 days of incubation), but it died when 6 days old.

With the Turkey

No authentic records of naturally occurring hybrids between the fowl and *Meleagris gallopavo* are known to the author. Warren and Scott (1935) and Quinn *et al.* (1937) tried to produce them by artificial insemination with little success. In the cross, turkey ♀ x fowl ♂, Quinn *et al.* incubated 656 eggs of which only 5 were fertile, and all of these died early. The reciprocal cross was more successful, the proportion of fertile eggs being about 20 per cent. A similar difference in fertility between the two crosses was noted by Warren and Scott. One of their hybrids lived to 22 days of incubation, and the other investigators had one survive a little longer.

Literature Cited

Asmundson, V. S. 1942. Crossbreeding and heterosis in turkeys. *Poultry Sci.,* 21:311-316.

Bice, C. M., and B. A. Tower. 1939. Crossbreeding poultry for meat production. *Hawaii Agr. Expt. Sta. Bull.* 81.

Bird, S., and J. W. Sinclair. 1938. On the validity of progeny tests of sires obtained on culled populations of daughters. *Sci. Agr.,* **19**:1-6.

Byerly, T. C., C. W. Knox, and M. A. Jull. 1934. Some genetic aspects of hatchability. *Poultry Sci.,* **13**:230-238.

Cole, L. J., and J. G. Halpin. 1916. Preliminary report of results of an experiment on close inbreeding in fowls. *J. Am. Assoc. Instruct. Invest. Poultry Husbandry,* 3:7-8.

— and —. 1922. Results of eight years of inbreeding of Rhode Island Red fowls. *Anat. Record,* **23**:97.

Cole, R. K., and F. B. Hutt. 1947. A genetic study of feathered feet in White Leghorns. *Poultry Sci.,* **26**:536.

Cutler, D. W. 1918. On the sterility of hybrids between the pheasant and the Gold Campine fowl. *J. Genetics,* 7:155-165.

Danforth, C. H., and G. Sandnes. 1939. Behavior of genes in intergeneric crosses. **J.** *Heredity,* **30**:537-542.

Dryden, J. 1921. Egg-laying characteristics of the hen. *Oregon Agr. Expt. Sta. Bull.* 180.

Dudley, F. J. 1934. Experiments on in-breeding poultry. *Ministry Agr. Fisheries, London, Bull.* 83.

—. 1944. Results of crossing the Rhode Island Red and White Leghorn breeds of poultry. *J. Agr. Sci.,* **34**:76-81.

Dumon, A. G. 1930. The effects of inbreeding on hatchability.

Proc. World's Poultry Congr., 11th Congr., London: 1-5.

—. 1938. La selection genealogique et consanguine de la poule pondeuse (Leghorn). *Agricultura (Louvain)*, **41**:11-41.

Dunkerly, J. S. 1930. The effect of inbreeding. *Proc. World's Poultry Congr., 4th Congr., London:* 47-72.

Dunn, L. C. 1923. Experiments on close inbreeding in fowls. *Storrs (Conn.) Agr. Expt. Sta. Bull.* 111.

—. 1928. The effect of inbreeding and crossbreeding on fowls. *Verhandl. V*

Intern. Kongr. Vererb., Z. ind. Abst. Vererb., Supplementb. **1**:609-617.

—. 1928a. The effect of inbreeding on the bones of the fowl. *Storrs (Conn.)* Agr. Expt. Sta. Bull, **152**:53-112.

Funk, E. M. 1934. Factors influencing hatchability in the domestic fowl. *Missouri Agr. Expt. Sta. Bull* 341.

—. 1938. The guinhen: hybrid resulting from crossing a guinea hen and a male chicken. *Am. Poultry J.,* **69**:16.

Ghigi, A. 1922. L'hybridisme dans la genese des races domestiques d'oiseaux. *Genetica,* **4**:364-374.

—. 1934. Richerche sull'origine delle razze domestiche di polli da forme selvatiche. Atti congr. mondiale pollicoltura. 5th Congr., Rome, **2**:3-13.

Godfrey, A. B. 1946. Value of the sisters' performance in selecting breeding cockerels. *Poultry Sci.,* **25**:148-156.

Guyer, M. F. 1909. Atavism in guinea-chicken hybrids. *J. Exptl. Zodl.,* **7**:723-745.

Guyer, M. F. 1912. Modifications in the testes of hybrids from the guinea and the common fowl. *J. MorphoL,* **23**:45-59.

Hall, G. 0. 1934. Breeding a low producing strain of Single Comb White Leghorns. *Poultry Sci.,* **13**:123-127.

—. 1935. The value of the pedigree in breeding for egg production. *Poultry Sci.,* **14**:323-329.

Hays, F. A. 1934. Effects of inbreeding on fecundity in Rhode Island Reds. *Mass. Agr. Expt. Sta. Bull.* 312.

—. 1940. Breeding small flocks of domestic fowl for high fecundity. *Poultry Sci.,* **19**:380-384.

Hess, C. W., T. C. Byerly, and M. A. Jull. 1941. The efficiency of feed utilization by Barred Plymouth Rock and crossbred broilers. *Poultry Sci.,* **20**:210-216.

Horlacher, W. R., R. M. Smith, and W. H. Wiley. 1941. Cross-breeding for broiler production. *Arkansas Agr. Expt. Sta. Bull.* 411.

Houwink, R. 1921. Wild fowls, their mutual relationship and their connection with domesticated breeds. *Trans. World's Poultry Congr. 1st Congr., The Hague-Scheveningen,* Appendix: 19-20.

Hutt, F. B., R. K. Cole, M. Ball, J. H. Bruckner, and R. F. Ball. 1944. A relation between environment to two weeks of age and mortality from lymphomatosis in adult fowls. *Poultry Sci,* **23**:396-404.

Jull, M. A. 1929. Studies in hatchability. II. Hatchability in relation to consanguinity of the breeding stock. *Poultry Sci.,* **8**:219-229.

—. 1929a. Studies in hatchability. III. Hatchability in relation to coefficients

of inbreeding. *Poultry Sci.,* 8:361-368.

—. 1930. Studies in hatchability. IV. The effect of intercrossing inbred strains of chickens on fertility and hatchability. *Poultry Sci.,* **9**:149-156.

—. 1933. Inbreeding and intercrossing in poultry. *J. Heredity,* **24**:93-101.

—. 1934. Limited value of ancestors' egg production in poultry breeding. J. Heredity, **25**:61-64.

Knox, C. W. 1946. The development and use of chicken inbreds. *Poultry Sci.,* **25**:262-272.

— and M. W. Olsen. 1938. A test of crossbred chickens. Single Comb White Leghorns and Rhode Island Reds. *Poultry Sci.,* **17**:193-199.

—, J. P. Quinn, and A. B. Godfrey. 1943. Comparison of Rhode Island Reds,

White Wyandottes, Light Sussex, and crosses among them to produce F1 and three-way cross progeny. *Poultry Sci.,* **22**:83-87.

Lamoreux, W. F., F. B. Hutt, and G. O. Hall. 1943. Breeding for low fecundity in the fowl with the aid of the progeny test. *Poultry*

Sci., **22**:161-169.

Lerner, I. M., and L. W. Taylor. 1940. The effect of controlled culling of chickens on the efficiency of progeny tests. *J. Agr. Research,* **60**:755-763.

Lippincott, W. A. 1920. Improving mongrel farm flocks through selected standard-bred cockerels. *Kansas Agr. Expt. Sta. Bull.* 223.

Lotsy, J. P., and K. Kuiper. 1924. A preliminary statement of the results of Mr. Houwink's experiments concerning the origin of some domestic animals. V. *Genetica,* **6**:221-277.

Marchlewski, J. 1937. Guinea-fowl (*Numida melcagris* L.) and common fowl *(Gallus domesticus* L.) hybrids obtained by means of artificial insemination. *Bull, intern, acad. polon. sci. Lettres, Classes sci. math. nat.* B, 11:127-131.

Maw, A. J. G. 1942. Crosses between inbred lines of the domestic fowl. *Poultry Sci.,* **21**:548-553.

Mehrhof, N. R., W. F. Ward, and o. K. Moore. 1945. Comparison of purebred and crossbred cockerels with respect to fattening and dressing qualities. *Florida Agr. Expt. Sta. Bull.* 410.

Mueller, C. D., and F. B. Hutt. 1946. The numbers of daughters necessary for progeny tests in the fowl. *Poultry Sci.* **25**:246-255.

Owen, R. D. 1941. Reciprocal crosses between the guinea and the domestic fowl. *J. Exptl. Zool,* **88**:187-217.

Pearl, R. 1911. Breeding poultry for egg production. *Maine Agr. Expt. Sta. Bull.,* **192**:113-176.

— and F. M. Surface. 1909. Data on the inheritance of fecundity obtained from the records of egg production of the daughters of "200-egg" hens. *Maine Agr. Expt. Sta. Bull,* **166**:49-84.

Petrov, S. G. 1935. Analysis of the development of high and low lines of Leghorns at Cornell. *Poultry Sci.,* **14**:330-339.

Poll, H. 1912. Mischlingsstudien VII. Mischlinge von Phasianus und Gallus. *Sitzber. Akad. Wiss. Berlin,* **38**:864-883.

Quinn, J. P., W. H. Burrows, and T. C. Byerly. 1937. Turkey-chicken hybrids. *J. Heredity,* **28**:169-173.

Sandnes, G. C., and W. Landauer. 1938. The sex ratio in the cross of *Phasianus torquatus* 9 X *Gallus domesticus Am. Naturalist,*

72:180-183.

Schmidt, J. 1922. Diallel crossings with the domestic fowl. *J. Genetics,* **12**:241-245.

Serebrovsky, A. S. 1929. Observations on interspecific hybrids of the fowl. *J. Genetics,* **21**:327-340.

Shoffner, R. N. 1948. The variation within an inbred line of S. C. W. Leghorns. *Poultry Sci.,* **27**:235-236.

Tiniakov, G. 1934. Peacock and hen hybrids, and a comparative analysis of the caryotype of their parents (Russian with English summary). *Biol. Zhur.,* **3**:41-63.

Trouessart, E. 1920. Hybrides de paon et de poule. *Rev. hist. nat. appliq.* (pt. **II**) **1**:100-102.

Viljoen, N. F., and J. H. de Bruin. 1935. Anatomical studies No. 59. On a false masculine hermaphrodite in an avian hybrid. *Onderstepoort J. Vet. Sci.,* **5**:351-356.

Warren, D. C. 1927. Hybrid vigor in poultry. *Poultry Sci.,* **7**:1-8.

—. 1942. The crossbreeding of poultry. *Kansas Agr. Expt. Sta. Bull.* 52.

— and H. M. Scott. 1935. An attempt to produce turkey-chicken hybrids. *J. Heredity,* 26:105-107.

Waters, N. F. 1938. The influence of inbred sires top-crossed on White Leghorn fowl. *Poultry Sci.,* **17**:490-497.

—. 1945. The influence of inbreeding on hatchability. *Poultry-Sci.,* **24**:329-334.

— and W. V. Lambert. 1936. Inbreeding in the White Leghorn fowl. *Iowa Agr. Expt. Sta. Research Bull.* 202.

— and -. 1936a. A ten year inbreeding experiment in the domestic fowl. *Poultry Sci.,* **15**:207-218.

Wright, S. 1923. Mendelian analysis of the pure breeds of livestock. I. The measurement of inbreeding and relationship. *J. Heredity,* **14**:339-348.

—. 1931. Evolution in Mendelian populations. *Genetics,* **16**:97-159.

Zorn, W., and H. F. Krallinger. 1934. Die Legeleistung von Bastarden zwischen weissen, einfachkammigen Leghorns und Lachshühnern. *Arch. Geflugelk.*, 8:233-250.

Symbols for the Genes of the Fowl

The study of many mutations in the past three decades has shed much light on the gene complex of the fowl but has created chaos with respect to the symbols by which the mutations have been designated. This is merely because numerous symbols have been assigned by various workers who were apparently unaware that these same symbols had been previously used for other mutations. The symbols *p* and *s* have each been used for no less than four different genes.

In the following list, the usual rules of priority have been followed to the best of the author's knowledge, and the authorities are given by whom the symbols were first used. Complete citations of their papers in which the symbols were introduced will be found in the literature lists of the chapters given in the last column. Symbols assigned for the first time in this book are starred [*]. For some of these, other symbols proposed earlier have had to be discarded merely because they had been used still earlier by someone else. It is hoped that the list will forestall similar troubles in the future.

Symbols have not been assigned to any variations that are not clearly caused by the action of one pair of genes. This excludes some characters apparently unifactorial but showing poor penetrance or much influence of modifying genes. It also excludes quantitative characters, such as egg production, age at first egg, etc.

In devising symbols, the author has been guided by the recommendations of the International Committee on Nomenclature for Genetics of the Mouse [*J. Heredity*, **31** (1940)]. It would be helpful in maintaining uniformity if those proposing new symbols for genes of the fowl would follow suit. The author will be grateful for information about any additions or corrections that should be made in the following list.

Table 10. Symbols for Genes of the Fowl

Sym-bol	Character		When introduced and by whom	Description, citation in Chap.
a	Albinism, autosomal	1933.	Warren, *J. Heredity*, **24**	7
ab	Barring, autosomal	*		7
al	Albinism, sex-linked	1941.	Mueller and Hutt, *J. Heredity,* **32**	7
Ap	Apterylosis	*		5
B	Barring, sex linked	1908.	Spillman, *Am. Natural-ist,* **42**	7
bd	Breda comb	*		4
Bl	Blue	1933.	Hertwig, *Verhandl. deut. zool.*	7
			Ges.	
Br	Brown eye	1933.	MacArthur, *Genetics,* **18**	14
By	Brachydactyly	*		3
c	Recessive white	1914.	Hadley, *Rhode Island Expt. Sta. Bull* 161	7
ch	Chondrodystrophy	1942.	Lamoreux, *J. Heredity,* **33**	3
Cl	Cornish lethal	*		3
cn	Crooked-neck dwarf	*		8
Cp	Creeper	1933.	Hutt, *Genetics,* **18**	3 and 14 †

Sym-bol	Character		When introduced and by whom	Description, citation in Chap.
Cr	Crest	1927.	Dunn and Jull, *J. Genetics*, **19**	5
D	Duplex comb	1923.	Punnett, "Heredity in Poultry"	4
dp	Diplopodia	*		3
dw	Dwarf	*		9
e	Columbian restriction	1918.	Lippincott, *Am. Naturalist*, **52**	7
F	Frizzling	1930.	Hutt, *J. Genetics*, **22**	5
Fl	Flightless	1940.	Hutt and Lamoreux, *J. Heredity*, **31**	5 and 14 †
fr	Fray	1938.	Warren, *J. Heredity*, **29**	5
g	Yellow head	1935.	Deakin and Robertson, *Am.*	6
			Naturalist, **69**	
h	Silkiness	1927.	Dunn and Jull, *J. Genetics*, **19**	5
Hf	Hen feathering	*		5
I	Dominant white	1913.	Hadley, *Rhode Island Expt. Sta. Bull.* 155	7
id	Dermal melanin	1936.	Hutt, *Neue Forschung. Tierzucht. u Abstammungsl.*	6
w	Cream	1932.	Taylor, *Proc. Intern. Genetic Congr.*, 6th *Congr.*	7

Sym-bol	Character		When introduced and by whom	Description, citation in Chap.
K	Slow feathering	1929.	Hertwig and Ritters-haus, *Z. ind. Abst. Ver-erb.*, **51**	5
ko	Head streak	1930.	Rittershaus, *ZiicMer.*, **2**	7
I	Recessive white lethal	1923.	Dunn, *Am. Naturalist*, **57**	8
la	Lacing	*		7
Li	Light down	1929.	Hertwig and Ritters-haus, *Z. ind. Abst. Ver-erb.*, **51**	7
to	Congenital loco	*		8
M	Multiple spurs	1941.	Hutt, *J. Heredity*, **32**	4
ma	Marbling	1933.	Hertwig, *Verhandl. deut. zool. Ges.*	7
Mb	Muffs and beard	*		5
md	Missing mandible	*		3
mf	Modified frizzling	1936.	Hutt, *J. Genetics*, **32**	5
mi	Microphthalmia	*		8
mo	Mottling	*		7
mx	Amaxilla	*		3
n	Naked	1938.	Hutt and Sturkie, *J. Heredity*, **29**	5
Na	Naked neck	1933.	Hertwig, *Verhandl. deut. zool. Ges.*	5

Sym-bol	Character		When introduced and by whom	Description, citation in Chap.
O	Blue egg	1940.	Hutt and Lamoreux, *J. Heredity,*, **31**	11 and 14 [†]
P	Pea comb	1906.	Bateson and Punnett, *Repts. Evol. Comm. Roy. Soc.,* **III**	4
pi	Pied	*		7
pk	Pink eye	*		7
Po	Polydactyly (fifth toe)	1927.	Dunn and Jull, *J. Genetics,* **19**	3
Po^d	Duplicate polydactyly	*		3
po	Normal toes	*		3
R	Rose comb	1906.	Bateson and Punnett, *Repts. Evol. Comm. Roy. Soc.,* **III**	4
Rp	Rumplessness	1936.	Dunn and Landauer,	
			J. Genetics, **33**	3
rp-2	Recessive rumplessness	1945.	Landauer, *Genetics,* **30**	3
rs	Red-splashed white	*		7
S	Silver	1917.	Goodale, *Science,* **46**	7 and 14 [†]
Sd	Sex-linked dilution	*		7
si	Spurless	*		4
sm	Short mandible	*		3
Sp	Spangling	1932.	Taylor, *J. Genetics,* 26	7
st	Unstriped	*		7

Sym-bol	Character		When introduced and by whom	Description, citation in Chap.
sy	Stickiness	*		8
su	Short upper beak	*		3
T	Normal (rapid) feathering	1946.	Jones and Hutt, *J. Heredity*, **37**	5
t^s	Retarded feathering	1946.	Jones and Hutt, *J. Heredity*, **37**	5
t	Tardy feathering	1946.	Jones and Hutt, *J. Heredity*, **37**	5
ta	Talpid	1942.	Cole, *J. Heredity*, **33**	8
td	Thyrogenous dwarfism	*		9
U	Uropygial	1936.	Hutt, *Neue Forschung. Tierzucht. u. Abstammungsl.*	4
v	Vulture hocks	*		5
W	White skin	1925.	Dunn. *Anat. Record*, **31**	6
wg	Wingless	*		3

* New symbols assigned for the first time in this book.

† When two chapters are cited, the first contains the description of the mutation, and the second cites the report in which the symbol was first used.

Glossary

The Glossary contains definitions of genetical terms used in the book and of a few others that the reader may encounter elsewhere.

Acrosome. The apical body at the anterior tip of the spermatozoon.

Allele. One of two alternative genes that have the same locus in homologous chromosomes or of the two contrasting characters induced by such genes; in multiple alleles, one of several. For example, the gene causing rose comb is the dominant allele of that for single comb.

Allelomorph. Synonymous with allele, and the original term introduced by Bateson in 1902; shortened to allele by some geneticists about 30 years later.

Anaphase. That stage of mitosis following the metaphase and in which the daughter chromosomes move toward the poles of the spindle.

Aneuploid. An adjective, frequently used as a noun, referring to an organism (or to part of one) that has more or less than (1) the complement of chromosomes normal for the species and sex or (2) some multiple of that complement. For examples, see trisomic and monosomic.

Atavism (atavistic). The reappearance of some ancestral character after a lapse of several generations in which it has not been evident.

Autosomal. Pertaining to an autosome or to characters induced by genes in one or more autosomes, in contradistinction to sex-linked characters.

Autosome. Any chromosome other than a sex chromosome.

Backcross. The mating of an F1 hybrid or of an equivalent heterozygote to one of the two parental varieties, types, or genotypes that produced the hybrid.

Bimodal. An adjective describing a distribution of frequencies that shows two classes as having greater numbers than others.

Binucleate. Having two nuclei.

Bivalent. Applied to homologous chromosomes when united or associated in pairs, especially during synapsis.

Blastoderm. A membrane formed in the blastodisc of the egg by repeated segmentation following fertilization.

Blastodisc. The germinal disc, or circular spot, at which the nucleus of the egg is located and in which segmentation begins after fertilization. By the time of laying (of hens' eggs) the blastodisc of a fertile egg is covered by the blastoderm, while that of an infertile one is disintegrating.

Carrier. An organism heterozygous with respect to some recessive character. While a bird of the genotype *Rr* is heterozygous for both rose and single combs, it would ordinarily be referred to as a carrier of the gene for single comb, which it does not show, but not as a carrier of the gene for rose comb, because that part of its genotype is revealed by the bird's rose comb.

Character. A convenient term, a shortened form of characteristic, used to designate any structure, trait, or function of an organism, whether it be hereditary or acquired. Genetic characters are said to be the end products of their causative genes, the net result of the interaction among genes and of genes with environmental influences.

Chromatid. One of the two genetically identical strands formed by a chromosome in anticipation of cell division.

Chromatin. The part of a cell's nucleus which forms the most conspicuous part of the nuclear network and of the chromosomes and is deeply stained by basic, or "nuclear," dyes.

Chromomeres. Chromatinic granules arranged in linear order and thus comprising the chromosome.

Chromosome map. A diagram showing what genes are known

to belong in specific linkage groups and their positions in relation to one another within those groups.

Chromosomes. Literally, colored bodies. When stained with basic dyes, they are visible under the microscope as rods, loops, or dots in dividing cells that have been properly "fixed." They carry the genes, arranged in linear order.

Cleavage. Division, or segmentation, of the fertilized egg; usually restricted to the earlier divisions when the numbers of cells resulting from the segmentations can still be counted.

Complementary genes. Genes which in combination cause an effect different from that of either gene by itself; for example, the genes *R* and *P,* which separately induce rose and pea combs, respectively, but in combination cause a walnut comb.

Congenital. Present at birth, but not necessarily of genetic orgin.

Coupling phase. That type of association of two linked pairs of genes in which the chromosome carrying them has either both dominant alleles or both recessive alleles; for example, *F I* and *f i. See also* repulsion phase.

Crossing over. The interchange (during meiosis) of segments of homologous chromosomes and thus of chains of genes in those segments; the term is also applied to the new combinations of characters that result from the interchange.

Crossover. When used as a noun in the geneticist's jargon, this refers to the organism which, because of crossing over during the formation of one (or both) of the gametes from which it arose, carries the alleles of two or more linked genes in a combination different from that in which they were transmitted to one (or both) of its parents. To illustrate, when a dihybrid that received *F* and *I* from one parent and *f* and *i* from the other produces by an *fi ii* mate some chicks that are *F* and *i* or *f* and *I,* these are crossovers, *i.e.,* new combinations in contrast with the parental combinations.

Crossover distance. The distance in a chromosome separating two linked genes, as measured by the amount of crossing over between them, and expressed as the proportion of gametes tested that give rise to crossovers.

Crossover unit. A measure of the distance between linked genes, and synonymous with the proportion of crossovers between them. To illustrate, if the amount of crossing over between F and I is 17 per cent, these two genes are said to be 17 crossover units apart.

Cryptomere. A genetic character that is present in an organism but rendered invisible by some other character; for example, color patterns in recessive white birds. Presence of the cryptomere can be demonstrated by breeding tests.

Cytology. The study of cells, their structure, properties, and functions.

Cytoplasm. The protoplasm of the cell exclusive of the nucleus.

Diakinesis. A stage of meiosis in which the homologous chromosomes come together in pairs prior to their separation.

Diallel crosses. The mating of two or more animals at different times to the same two or more animals of the opposite sex in order thus to evaluate their genotypes with respect to some quantitative character; one kind of progeny test.

Dihybrid. Heterozygous with respect to two pairs of genes.

Diploid. Having two sets of chromosomes, as in somatic cells of animals, in contradistinction to the haploid state of the germ cells, which have only a single set of chromosomes.

Dominant. Referring to genes or characters that are manifested by organisms heterozygous for them, in contradistinction to recessive genes or characters, which are revealed only in homozygotes.

Duplicate genes. Two pairs of genes with equal effects on one character.

Dysgenic. Conducive to the accumulation of undesirable genes in the germ plasm of a species or population and hence to the weakening of the racial fitness of future generations; the opposite of eugenic.

Egg. A female reproductive cell. Strictly defined, a hen's oocyte becomes an egg when it extrudes the second polar body. This occurs soon after the egg enters the oviduct, or just before, and hence about 24 to 30 hours before laying.

Epistasis. The masking of the action of one gene (or pair of genes) by another that is not allelomorphic to the first. For example, sex-linked barring is dominant to non-barring, but the barring carried by White Leghorns does not show because the dominant white of that breed prevents the formation of color. Dominant white is thus epistatic to color and to barring.

Epistatic. *See* epistasis.

Equatorial plate. The plate formed by the chromosomes as they lie at the equator of the spindle in the metaphase stage of mitosis.

Estrogen. A feminizing substance that may be produced by the ovary, by other organs, or synthetically, but having in all cases the physiological properties of the normal ovarian hormone.

Eugenic. Conducive to the improvement of "the racial qualities of future generations, either physically or mentally" (Galton).

F1 generation. The first filial generation; the first generation from some specific mating.

F2 generation. The second filial generation from some specific mating. It is produced by mating the F1 generation *inter se* or by its self-fertilization in species in which that is possible.

Factor. Same as gene.

Fecundity. Although in most biological writing this refers to the capacity for reproduction and is often used synonymously with fertility, it has a special connotation when applied to domestic birds. In this narrower sense it refers to the capacity for laying eggs, whether these be fertile or infertile, and whether used for reproduction or not.

Fertilization. Fusion of the male and female pronuclei, following penetration of the egg by the male gamete, thus reestablishing the diploid number of chromosomes and initiating development of the embryo.

Gamete. A reproductive cell, whether an egg or a sperm.

Gametogenesis. The formation of gametes.

Gene. A unit of inheritance, carried in a chromosome and transmitted in the germ cells, which, by interacting with other genes, with the cytoplasm, and with the environment, influences the development of a character and sometimes controls it completely.

Genetics. The science that deals with heredity and variation, seeking to elucidate the principles underlying the former and the causes for the latter.

Genotype. The genetic constitution of an organism, including genes without visible effect as well as those revealed by the phenotype. It may refer to all the genes or to a single pair.

Germ cell. A reproductive cell, or one capable of giving rise to gametes, in contradistinction to somatic, or body, cells, which cannot do so.

Gonad. A reproductive gland; an ovary, testis, or ovotestis.

Gonadotropic. An adjective usually applied to a hormone or hormonal substance that is capable of activating or stimulating the gonad.

Gynandromorph. An animal in which part of the body is genetically female and another part genetically male.

Haploid. Single; referring to the reduced number of chromosomes as found in the gametes. *See also* diploid.

Hemizygous. Referring to the single state of sex-linked genes in individuals of the heterogametic sex. To illustrate, a Barred Rock male may be homozygous or heterozygous for barring, but the female can be neither, as she has only one sex chromosome. She must be hemizygous, if barred.

Hermaphrodite. An organism that produces both eggs and sperms; sometimes erroneously applied to fowls that are hormonal intersexes.

Heterogametic. Producing gametes of two kinds with respect to sex chromosomes, the difference being such that one kind induces (upon fertilization) a male zygote and the other a female.

Heteroploid. Referring to a complement of chromosomes that is not a multiple of the haploid number.

Heterosis. Synonymous with hybrid vigor.

Heterozygote. The noun applied to an organism that is heterozygous for some specific pair or pairs of genes and hence produces gametes of two kinds with respect to the alleles for which it is heterozygous.

Heterozygous. Carrying both the dominant and the recessive genes of a pair of alleles or two different genes of a series of multiple alleles.

Homogametic. Producing gametes that are all of equal potency with respect to the sex of the zygote to which they give rise upon fertilization.

Homologous chromosomes. Those occurring in somatic cells in pairs, both being alike in size and form and differing from each other with respect to genes less than they differ from other pairs. One member of a pair is normally inherited from each parent.

Homozygote. The noun applied to an organism homozygous for some specific pair or pairs of genes and hence producing gametes all alike with respect to the genes concerned.

Homozygous. Carrying two of either the dominant or recessive genes of a pair of alleles, or carrying two identical genes of a series of multiple alleles.

Hormone. The biologically active secretion of a ductless gland.

Hybrid. In general biological usage, this refers to the offspring of a cross between different species or varieties. Geneticists apply it also to organisms heterozygous with respect to one or more pairs of genes. Thus, when Rose-comb and Single-comb Brown Leghorns are crossed, the rose comb is said to be dominant in the *hybrid,* even though these hybrids are phenotypically indistinguishable from pure Rose-comb Brown Leghorns.

Hybrid vigor. The extra vigor, exceeding that of their parent stocks, frequently shown by the hybrids from the crossing of such genetically dissimilar parents as

different species, breeds, strains, or inbred lines. It may be expressed as more rapid growth, larger size, greater viability, or otherwise.

Hypostatic. The adjective applied to a character that is masked by some recessive character not an allele of the first. To illustrate, sex-linked barring is hypostatic to recessive white, so that White Plymouth Rocks cannot show the barring that most of them carry. Recessive white is thus epistatic to barring.

Inbreeding. The mating of relatives.

Inbreeding, coefficient of. A measure of the extent to which inbreeding has reduced heterozygosity in comparison with that prevailing in similar animals not inbred.

Inhibitor. A gene or character suppressing some process or function that would proceed if the inhibitor were not present. For example, the gene *Id* inhibits the deposition of melanin in the dermis and hence induces white or yellow color in shanks that would otherwise be blue, slate, or willow.

Inter se. Among themselves.

Intersex. An organism showing characteristics intermediate between those of normal males and females of its species.

Lethal gene. A gene that at some time between fertilization of the egg and the normal life span of the species causes the premature death of the organism homozygous or hemizygous for it; in a narrower sense, often restricted to genes causing death before birth, at birth, or soon after it.

Line breeding. A form of inbreeding that concentrates the blood (genes) of some specific ancestor, *i.e.,* the mating of later generations back to that ancestor or to its descendants.

Linkage. The association of two or more characters in inheritance because their causative genes are located in the same chromosome.

Linkage group. A group of genetic characters each of which has been shown by suitable tests to be linked with other members of the same group. The number of linkage groups in a species cannot exceed the number of pairs of chromosomes that is normal for that species.

Locus. The position of a gene in a chromosome or in a linkage group, usually stated in terms of its relation to other genes in the same chromosome.

Loose linkage. A term used somewhat loosely in reference to genes that are linked but so far apart on the chromosome that crossing over between them may be of the order of 30 per cent or more.

Mass selection. Selection of breeding stock based on the appearance or performance of individuals, *i.e.,* phenotypic selection.

Matroclinous. Resembling the mother.

Maturation divisions. Same as meiosis.

Meiosis. That process (in the formation of gametes) consisting of two cell divisions by which the number of chromosomes is reduced in the gametes to half the number found in somatic cells.

Metaphase. The stage of cell division at which the chromosomes are arranged in the equatorial plate near the centre of the cell.

Micron *(μ).* The one-thousandth part of a millimetre; a measure commonly used in the microscopical study of cells and tissues.

Mitosis. The normal type of cell division in somatic cells, during which the chromosomes become condensed, duplicated (by a process usually said to be division), aligned at the centre of the cell, and thence carried to the two daughter cells in such a way that both these receive identical chromosomes.

Modifier, or modifying gene. A gene, usually with lesser effects, that influences the expression of some other gene.

Monohybrid. Heterozygous with respect to a single pair of genes.

Monosomic. Having one chromosome less than the normal diploid number for the species and sex concerned.

Morphological character. A genetic variation in form or structure, in contradistinction to physiological characters, which are variations in function. The distinction is often arbitrary.

Morphology. The study of form and structure.

Mosaic. An organism in which some part of the body has one genotype and the remaining part has another.

Multifactorial. Referring to a character that is dependent for its expression upon the combined action of an undetermined number of genes.

Multiple alleles. A series of three or more genes any one of which may occupy a single locus on a chromosome. All of them influence the same character, but in differing degrees.

Multiple factors. An undetermined number of genes that together influence the expression of some one genetic character.

Mutation. A sudden change in a gene or in the chromosomes, resulting in a new variation that is hereditary. It may refer either to the invisible change in the cells or to the resultant visible change in the chicken.

New combinations. Applied in tests for linkage to the combinations of genes or characters that differ from the parental combinations of the genes concerned. If the genes are linked, the new combinations are also crossovers, but otherwise they result merely from independent assortment.

Non-disjunction. The failure of homologous chromosomes to separate during meiosis, so that one daughter cell gets both, the other one neither.

Normal overlaps. In a population showing continuous intergradation between some mutant character and the normal type, those members which are indistinguishable from normal ones though carrying the genotype which in most individuals causes some expression of the character.

Nucleus. A small body within a cell that stains deeply with basic dyes, contains the chromosomes, and reproduces itself in cell division.

Oocyte. An egg cell prior to completion of the process of maturation, *i.e.,* prior to extrusion of the second polar body.

Oogenesis. Formation of the egg, referring particularly to the process of meiosis, or maturation, by which the primary oocyte becomes an egg with the haploid number of chromosomes.

Oogonia. Cells in the ovary which are descended from the primordial germ cells and of which some eventually become oocytes.

Outcross. The mating of stock that is somewhat inbred or homogeneous to unrelated individuals.

Ovariotomy. Removal of the ovary; synonymous wtih ovariectomy.

Ovotestis. A single gonad of which part resembles an ovary in its histological structure and part resembles a testis.

Ovulation. Escape of the egg from the ovarian follicle; not the laying of the egg, which is termed "oviposition."

Ovum. An egg. Strictly speaking, it refers to an egg at the stage when it is ready for fertilization.

Parental combinations. Applied in tests for linkage to the particular combinations of genes or characters that were present in the two parents that produced the dihybrid (or polyhybrid) individual used for the test; sometimes called "non-crossover" combinations.

Patroclinous. Resembling the father.

Penetrance. The extent to which a character is expressed in a group of individuals homozygous for it; measured as the proportion (per cent) of such individuals that show the character.

Phenotype. The genetic nature of an organism in so far as it is revealed by visible characters or measurable performance, in contradistinction to the genotype, which may not be evident without a breeding test.

Physiological character. A hereditary character affecting function rather than form or structure. *See also* morphological character. To illustrate, frizzling of the feathers might be considered a morphological character, but the rapid growth of feathers induced by the gene k is a physiological character.

Polar body. One of two minute cells containing a haploid set of chromosomes that are extruded by the ovum before fusion of the male and female pronuclei.

Polychromatism. Occurring in several different colors.

Polyploid. An organism having three or more complete sets of homologous chromosomes, *i.e.,* at least one set in excess of the normal diploid number.

Polyspermy. Penetration of the egg by more than one sperm, a condition that is normal in the eggs of birds.

Poularde. An ovariotomized female fowl, or one lacking female hormones because the ovary is undeveloped.

Progeny test. Evaluation of the genotype of an animal by the kinds of offspring that it produces.

Pronucleus. The nucleus of an egg or sperm during fertilization.

Prophase. The stage in cell division from the first appearance of the chromosomes up to the metaphase.

Protoplasm. The active or living substance of the cell, including the nucleus and the cytoplasm.

Proven sire. One with enough progeny that have been measured (in appearance or in performance) to reveal the kind of inheritance transmitted to them, whether good or bad, and hence to give some idea of the genotype of the sire.

Pseudogynandromorph. An animal appearing to be male in one part and female in another, but really of one sex throughout.

Quantitative character. An inherited character the expression of which is influenced by multiple factors in such a way that there is continuous intergradation between the extremes of its expression; a multifactorial character; for examples, fecundity, body size.

Recessive. An adjective, frequently used as a noun, referring to a gene or character that is expressed only by homozygotes; that member of a pair of alleles which does not show in heterozygotes.

Reciprocal crosses. Two crosses between two species, breeds, strains, or genotypes, A and B, such that one mating is $\male\ A \times \female\ B$, and the other is $\male\ B \times \female\ A$.

Recombinations. In tests for linkage, the progeny that show a combination of characters different from either of the parental combinations; crossovers; also new combinations that are not crossovers.

Repulsion phase. That type of association of two linked pairs of genes in which the chromosome carrying them has the dominant allele of one pair and the recessive allele of the other; for example, $F\,i$ and $f\,I$. *See also* coupling phase.

Reversion. The appearance of an individual that differs in some respect from both its parents but resembles a grandparent or some ancestor more remote; frequently used with reference to domestic animals that resemble the wild-type ancestor from which the domestic forms have arisen.

Secondary sexual character. A character induced by hormones of the ovary or testis, but not including variations in the reproductive tract, which are sometimes designated as primary sexual characters.

Segregation. The separation, during meiosis, of allelomorphic genes and the random recombination at fertilization of the two kinds of gametes thus produced, with the result that different types appear in the progeny in typical Mendelian ratios.

Semi-lethal. In genetic parlance this term implies that a gene or character is lethal to part of the population having the genotype for its expression, and *not* that it leaves any individuals half dead.

Sex chromosomes. Chromosomes that exert a preponderant influence upon the determination of sex. In animals the homogametic sex has two homologous sex chromosomes, and the heterogametic sex has either one of these alone or one with a dissimilar mate.

Sex dichromatism. Different colors in the two sexes.

Sex dimorphism. A difference between the sexes in size, structure, color, or some other attribute. •

Sex limited. Manifested only in one sex, *e.g.*, egg production.

Sex linked. An adjective applied to a gene carried in the kind of sex chromosome that is paired in the homogametic sex or to the character induced by such a gene.

Sex-linked cross. A cross between individuals carrying different alleles of a pair of sex-linked genes in such a way that the character shown by one sex appears among the offspring in the opposite sex only. This occurs when the heterogametic sex carries the dominant allele, as in the cross in the fowl, barred ♀ x non-barred ♂.

Siblings. Brothers and sisters.

Sib-tested. This adjective is applied to an individual for which the genotype with respect to some quantitative character has been estimated from the phenotypes of its brothers or sisters, or both.

Simple. In addition to its usual meaning, this word has a special connotation in . the geneticist's vocabulary, being applied to characters dependent for their expression upon the action of a single pair of genes.

Somatic. Referring to all cells and tissues other than germ cells.

Sperm. A shortened form of the word "spermatozoon" or of its plural form.

Spermatid. One of the four cells derived from a primary spermatocyte by the meiotic divisions and which, with some modification in shape but with no further cell divisions, becomes a spermatozoon.

Spermatocyte. A male reproductive cell prior to completion of the maturation divisions. Primary spermatocytes are derived from spermatogonia, and the first division of a primary spermatocyte produces two secondary ones.

Spermatogenesis. The formation of spermatozoa.

Spermatogonia. Male germ cells at stages intermediate between the primordial germ cell and the primary spermatocyte.

Spermatozoon (pi. -zoa). A mature male germ cell.

Spindle. A group of structures which in fixed preparations of dividing cells resemble threads arranged in the form of a spindle.

Sport. A mutation, referring particularly to the visible character and not to the change in the chromosomes.

Symbol. One or two letters, sometimes more, or sometimes a combination of letters and numbers, used to designate a gene or character so that its name need not be written out in full.

Synapsis. The union in pairs of homologous chromosomes (one of maternal and one of paternal origin) to form bivalent chromosomes during one stage of meiosis.

Telophase. The last stage of mitosis, when movement of the chromosomes has ceased.

Tetrad. The quadruple group of chromatids formed during meiosis by the association of two homologous chromosomes each of which consists of two chromatids.

Transgressive inheritance. The appearance in an F2 generation (or later) of individuals showing a more extreme development of some - character than that seen in the original parents that were crossed.

Trisomic. Having one chromosome more than the normal diploid number for the species and sex concerned.

Unifactorial. An adjective applied to a character that is dependent for its expression upon the action of a single pair of genes. Such a character may be dominant, in which case it is expressed in the heterozygote, or recessive, in which case it is revealed only by recessive homozygotes.

Unit character. Same as unifactorial character.

W chromosome. For species having heterogametic females (as in birds and the Lepidoptera) this term is sometimes used to designate the unpaired sex chromosome which occurs only in females. It is the counterpart of the Y chromosome. It is doubtful whether the use of Z and W introduces clarity or confusion. It would probably be simpler to use only X and Y, as one must know in either case whether the species concerned has heterogametic males or females.

Wild type. The normal phenotype of a species as it occurs in nature or (in the case of some domesticated species) as it is presumed to have been prior to domestication. It is a convenient norm by comparison with which mutations may be designated as dominant or recessive to the wild type.

X chromosome. A designation commonly used for the sex chromosome which, in species having heterogametic males (as in mammals), occurs singly in the male but is paired in the female. It can also be applied to the corresponding chromosome in species having the female heterogametic. *See also Z* chromosome.

Xenia. The immediate effect of male gametes (after mating) upon some maternally determined character. It occurs in some plants but has not been demonstrated in animals.

Y chromosome. The sex chromosome that occurs singly in heterogametic males or females and does not occur in the opposite sex of the same species. *See also W* chromosome.

Z chromosome. A designation sometimes used, in referring to species having heterogametic females, for the sex chromosome which occurs singly in such females but is paired in males of the same species. It is thus commonly used in writing of birds, moths, and butterflies by those who prefer to restrict the homologous term, X chromosome, to the type of sex determination in which the male is heterogametic. *See also W* chromosome.

Zygote. The cell formed by union of male and female gametes; the fertilized egg.

www.ingramcontent.com/pod-product-compliance
Lightning Source LLC
Chambersburg PA
CBHW031951190326
41519CB00007B/763